Perspectives on Tannins

Perspectives on Tannins

Editor
Andrzej Szczurek

MDPI • Basel • Beijing • Wuhan • Barcelona • Belgrade • Manchester • Tokyo • Cluj • Tianjin

Editor
Andrzej Szczurek
Centre of New Technologies,
University of Warsaw
Poland

Editorial Office
MDPI
St. Alban-Anlage 66
4052 Basel, Switzerland

This is a reprint of articles from the Special Issue published online in the open access journal *Biomolecules* (ISSN 2218-273X) (available at: https://www.mdpi.com/journal/biomolecules/special_issues/Perspectives_Tannins).

For citation purposes, cite each article independently as indicated on the article page online and as indicated below:

LastName, A.A.; LastName, B.B.; LastName, C.C. Article Title. *Journal Name* **Year**, *Volume Number*, Page Range.

ISBN 978-3-0365-1092-7 (Hbk)
ISBN 978-3-0365-1093-4 (PDF)

© 2021 by the authors. Articles in this book are Open Access and distributed under the Creative Commons Attribution (CC BY) license, which allows users to download, copy and build upon published articles, as long as the author and publisher are properly credited, which ensures maximum dissemination and a wider impact of our publications.

The book as a whole is distributed by MDPI under the terms and conditions of the Creative Commons license CC BY-NC-ND.

Contents

About the Editor vii

Andrzej Szczurek
Perspectives on Tannins
Reprinted from: *Biomolecules* **2021**, *11*, 442, doi:10.3390/biom11030442 1

Xiaowei Sun, Haley N. Ferguson and Ann E. Hagerman
Conformation and Aggregation of Human Serum Albumin in the Presence of Green Tea Polyphenol (EGCg) and/or Palmitic Acid
Reprinted from: *Biomolecules* **2019**, *9*, 705, doi:10.3390/biom9110705 5

Jing Chen, W. P. D. Wass Thilakarathna, Tessema Astatkie and H. P. Vasantha Rupasinghe
Optimization of Catechin and Proanthocyanidin Recovery from Grape Seeds Using Microwave-Assisted Extraction
Reprinted from: *Biomolecules* **2020**, *10*, 243, doi:10.3390/biom10020243 19

Ling Guo, Zhi-Yi Yang, Ren-Cheng Tang and Hua-Bin Yuan
Preliminary Studies on the Application of Grape Seed Extract in the Dyeing and Functional Modification of Cotton Fabric
Reprinted from: *Biomolecules* **2020**, *10*, 220, doi:10.3390/biom10020220 35

Anna Grobelna, Stanisław Kalisz and Marek Kieliszek
The Effect of the Addition of Blue Honeysuckle Berry Juice to Apple Juice on the Selected Quality Characteristics, Anthocyanin Stability, and Antioxidant Properties
Reprinted from: *Biomolecules* **2019**, *9*, 744, doi:10.3390/biom9110744 47

Thomas Sepperer, Gianluca Tondi, Alexander Petutschnigg, Timothy M. Young and Konrad Steiner
Mitigation of Ammonia Emissions from Cattle Manure Slurry by Tannins and Tannin-Based Polymers
Reprinted from: *Biomolecules* **2020**, *10*, 581, doi:10.3390/biom10040581 61

Flavia Lega Braghiroli, Gisele Amaral-Labat, Alan Fernando Ney Boss, Clément Lacoste and Antonio Pizzi
Tannin Gels and Their Carbon Derivatives: A Review
Reprinted from: *Biomolecules* **2019**, *9*, 587, doi:10.3390/biom9100587 73

Antonio Pizzi
Tannins: Prospectives and Actual Industrial Applications
Reprinted from: *Biomolecules* **2019**, *9*, 344, doi:10.3390/biom9080344 107

About the Editor

Andrzej Szczurek linked his scientific carrier with carbonaceous materials and their properties. He received his Ph.D. degree in Material Chemistry in 2011 from the University of Lorraine (France). Since 2017, he has been an assistant professor in the Centre of New Technologies, at the University of Warsaw. His research interests touch on multiple topics, such as tannin and lignin-based polymeric and carbon materials, or the new materials, e.g., graphynes and graphdiynes chemistry. He is interested in materials properties, particularly gas adsorption, as well as the electrochemical and catalytic activity of porous materials.

Editorial

Perspectives on Tannins

Andrzej Szczurek

Centre of New Technologies, University of Warsaw, S. Banacha 2C, 02097 Warsaw, Poland; a.szczurek@cent.uw.edu.pl

Citation: Szczurek, A. Perspectives on Tannins. *Biomolecules* **2021**, *11*, 442.

https://doi.org/10.3390/biom11030442

Received: 12 March 2021
Accepted: 15 March 2021
Published: 16 March 2021

Publisher's Note: MDPI stays neutral with regard to jurisdictional claims in published maps and institutional affiliations.

Copyright: © 2021 by the author. Licensee MDPI, Basel, Switzerland. This article is an open access article distributed under the terms and conditions of the Creative Commons Attribution (CC BY) license (https://creativecommons.org/licenses/by/4.0/).

Tannins are a family of versatile, natural phenolic biomolecules whose main role is to protect plants against insects and fungi. Although this group encompasses a wide variety of oligomers and polymers, two main categories of tannins can be distinguished: hydrolysable tannins (HTs) and proanthocyanidins (Pas), known also as condensed flavonoid tannins, which are resistant to hydrolytic degradation [1]. Tannins contain aromatic rings bearing hydroxyl groups, which give them high chemical activity, causing them to form complexes with other macromolecules, such as carbohydrates [2], or bacterial cell membranes [3]. However, their main characteristics are complexation and precipitation of proteins [4]. Tannins exhibit high antioxidant properties applicable to the food and medical industries [5]. It has been shown that they prevent oxidative-stress-related diseases, such as cardiovascular disease, cancer, and osteoporosis. Due to their chemical reactivity, high availability in raw materials, and easy and safe extraction, tannins are widely used in the food, leather, and chemical industries [1]. Examples of current or potential applications include their use in the production of anticorrosive primers [6], as "green" alternatives to synthetic homologous phenolic compounds for the production of wood adhesives and particle boards [7], and in functional polymeric materials with variable physicochemical and morphological properties [8]. In addition, the pyrolysis of these materials, i.e., heat treatment in inert atmosphere at elevated temperatures, results in cellular carbonaceous materials [9], further broadening the range of possible applications.

This book includes seven papers: five original articles and two review articles describing the versatile properties and applications of tannin biomolecules and tannin-based materials.

Sun et al. [10] investigated and described the role of epigallocatechin gallate (EGCg) in the conformation and agglomeration of human serum albumin (HSA). Their research was conducted in the presence or not of palmitic acid (PA). It was shown that EGCg increased the interdomain distance in HSA and HSA-PA. Regarding the effect of palmitic acid, the distance depended on the PA concentration. The EGCg also affected the secondary structure of the protein more significantly than palmitic acid. The EGCg decreased the α-helix content in a dose-dependent fashion. It was able to increase the HSA aggregation, whereas it promoted the formation of more heterogeneous aggregates. Any of these effects could impact the ability of serum albumin to transport and stabilize ligands, including EGCg and other polyphenols.

The article by Chen et al. [11] dealt with the applicability of microwave-assisted extraction to convert high molecular weight proanthocyanidins into monomeric catechins. The MSE extraction resulted in higher yields of monomeric catechins and PACs in comparison with conventional methods. Furthermore, the obtained extracts presented higher antioxidant capacity and α-glucosidase inhibitory activity than those prepared in conventional extractions. This finding suggests the potential use of the MAE products of grape seeds as functional food ingredients and nutraceuticals.

Guo et al. [12] revealed that proanthocyanins extracted from grape seeds might be able to serve as the effective dyes used to color cotton fabrics. They proved that proanthocyanins provided antibacterial, antioxidant, and UV protection to the treated cotton fabrics.

The study by Grobelna et al. [13] showed the higher stability and antioxidant activity of apple juice mixed with blue honeysuckle berry juice. The results were strictly due to the addition of the blue honeysuckle berry juice, which is rich in phenolic compounds (mostly anthocyanin).

Sepperer et al. [14] explored the capability of different types of tannins and tannin-based materials to capture ammonia. Industrial livestock farming produces enormous amounts of cattle manure, a source of ammonia. This is an important environmental issue due to the impacts on soil and air, and the European community has committed to reducing ammonia emissions by 30% as compared to 2005 by 2030. In their study, the ammonia adsorption capabilities of different types of tannins and tannin-based materials were considered. The obtained results were reasonably promising, with tannins showing a high ability to adsorb NH_3 and reduce pH. These results open up new possibilities for the use of materials produced from tannins.

Braghiroli et al. [15] extensively reviewed the recent achievements related to the preparation of tannin-based gels and their carbon derivatives. In this review, all crucial aspects of gels preparation (synthesis, drying, and pyrolysis conditions) regarding their final properties were recalled and discussed. The review also provided examples of the practical application of received gels for energy storage, thermal insulation, and contaminant sorption in drinking water and wastewater.

Pizzi [16] reviewed known and established applications of tannin-based materials, as well as the future applications that are being developed at present and that promise to have an industrial impact in the future. The essential points of the materials, their drawbacks, and their likelihood of industrial application were briefly discussed. The chemical nature of these applications was described, for which it is essential to understand the roles of tannins and their derivatives.

This book, titled "Perspectives on Tannins", presents articles and reviews disclosing the most relevant discoveries related to the antioxidant, antibacterial, and UV protection features of tannins. Polymerized tannin materials may find applications as adsorbents of toxic by-products. The chemical activity of tannins, their variety of forms, and their versatile properties make them suitable for a very wide range of applications.

Acknowledgments: I would like to thank all authors for their valuable contributions to this Special Issue and all reviewers for their helpful comments during the peer review process.

Conflicts of Interest: The author declares no conflict of interest.

References

1. Quideau, S.; Deffieux, D.; Douat-Casassus, C.; Pouysgu, L. Plant Polyphenols: Chemical Properties, Biological Activities, and Synthesis. *Angew. Chem. Int. Ed.* **2011**, *50*, 586–621. [CrossRef]
2. Amoako, D.; Awika, J.M. Polyphenol interaction with food carbohydrates and consequences on availability of dietary glucose. *Curr. Opin. Food Sci.* **2016**, *8*, 14–18. [CrossRef]
3. Smith, A.H.; Zoethendal, E.; Mackie, R. Bacterial Mechanisms to Overcome Inhibitory Effects of Dietary Tannins. *Microb. Ecol.* **2005**, *50*, 197–205. [CrossRef] [PubMed]
4. Adamczyk, B.; Salminen, J.-P.; Smolander, A.; Kitunen, V. Precipitation of proteins by tannins: Effects of concentration, protein/tannin ratio and pH. *J. Food Sci. Technol.* **2012**, *47*, 875–878. [CrossRef]
5. Chung, K.-T.; Wong, T.Y.; Wei, C.-I.; Huang, Y.-W.; Lin, Y. Tannins and Human Health: A Review. *Crit. Rev. Food Sci.* **1998**, *38*, 421–464. [CrossRef] [PubMed]
6. Hornus Sack, S.; Romagnoli, R.; Vetere, V.F.; Eisner, C.I.; Pardini, O.; Arnalvy, J.I.; Di Sarli, A.R. Evaluation of Steel/Primer Based On Chestnut Tannin/Paint Film Systems by EIS. *J. Coat. Technol.* **2002**, *74*, 63–69. [CrossRef]
7. Matsumae, T.; Horito, M.; Kurushima, N.; Yoshikazu Yazaki, Y. Development of bark-based adhesives for plywood: Utilization of flavonoid compounds from bark and wood. II. *J. Wood Sci.* **2019**, *65*, 9. [CrossRef]
8. Celzard, A.; Szczurek, A.; Prasanta, J.; Fierro, V.; Basso, M.-C.; Bourbigot, S.; Stauber, M.; Pizzi, A. Latest progresses in the preparation of tannin-based cellular solids. *J. Cell. Plast.* **2015**, *51*, 89–102. [CrossRef]
9. Szczurek, A.; Fierro, V.; Pizzi, A.; Celzard, A. Mayonnaise, whipped cream and meringue, a new carbon cuisine. *Carbon* **2013**, *58*, 245–248. [CrossRef]
10. Sun, X.; Ferguson, H.N.; Hagerman, A.E. Conformation and Aggregation of Human Serum Albumin in the Presence of Green Tea Polyphenol (EGCg) and/or Palmitic Acid. *Biomolecules* **2019**, *9*, 705. [CrossRef] [PubMed]
11. Chen, J.; Thilakarathna, W.P.D.W.; Astatkie, T.; Rupasinghe, H.P.V. Optimization of Catechin and Proanthocyanidin Recovery from Grape Seeds Using Microwave-Assisted Extraction. *Biomolecules* **2020**, *10*, 243. [CrossRef] [PubMed]

12. Guo, L.; Yang, Z.-Y.; Tang, R.-C.; Yuan, H.-B. Preliminary Studies on the Application of Grape Seed Extract in the Dyeing and Functional Modification of Cotton Fabric. *Biomolecules* **2020**, *10*, 220. [CrossRef] [PubMed]
13. Grobelna, A.; Kalisz, S.; Kieliszek, M. The Effect of the Addition of Blue Honeysuckle Berry Juice to Apple Juice on the Selected Quality Characteristics, Anthocyanin Stability, and Antioxidant Properties. *Biomolecules* **2019**, *9*, 744. [CrossRef] [PubMed]
14. Sepperer, T.; Tondi, G.; Petutschnigg, A.; Young, T.M.; Steiner, K. Mitigation of Ammonia Emissions from Cattle Manure Slurry by Tannins and Tannin-Based Polymers. *Biomolecules* **2020**, *10*, 581. [CrossRef] [PubMed]
15. Braghiroli, F.L.; Amaral-Labat, G.; Boss, A.F.N.; Lacoste, C.; Pizzi, A. Tannin Gels and Their Carbon Derivatives: A Review. *Biomolecules* **2019**, *9*, 587. [CrossRef] [PubMed]
16. Pizzi, A. Tannins: Prospectives and Actual Industrial Applications. *Biomolecules* **2019**, *9*, 344. [CrossRef] [PubMed]

Article

Conformation and Aggregation of Human Serum Albumin in the Presence of Green Tea Polyphenol (EGCg) and/or Palmitic Acid

Xiaowei Sun, Haley N. Ferguson and Ann E. Hagerman *

Department of Chemistry & Biochemistry, Miami University, Oxford, OH 45056, USA; sun.2766@osu.edu (X.S.); fergush2@miamioh.edu (H.N.F.)
* Correspondence: hagermae@miamioh.edu; Tel.: +1-513-529-2827

Received: 7 October 2019; Accepted: 2 November 2019; Published: 5 November 2019

Abstract: Polyphenols such as epigallocatechin gallate (EGCg) may have roles in preventing some chronic diseases when they are ingested as components of plant-based foods and beverages. Human serum albumin (HSA) is a multi-domain protein that binds various ligands and aids in their transport, distribution, and metabolism in the circulatory system. In the present study, the HSA-EGCg interaction in the absence or presence of fatty acid has been investigated. Förster resonance energy transfer (FRET) was used to determine inter- and intra-domain distances in the protein with and without EGCg and palmitic acid (PA). By labeling Cys-34 with 7-(diethyl amino)-4-methylcoumarin 3-maleimide (CPM), the distance between Trp-214 at domain IIA and CPM-Cys-34 at domain IA could be established. A small amount of PA decreased the distance, while a large amount increased the distance up to 5.4 Å. EGCg increased the inter-domain distance in HSA and HSA-PA up to 2.8 and 7.6 Å, respectively. We concluded that PA affects protein conformation more significantly compared to EGCg. Circular dichroism (CD) established that EGCg affects protein secondary structure more significantly than PA. PA had little effect on the α-helix content of HSA, while EGCg decreased the α-helix content in a dose-dependent fashion. Moreover, EGCg decreased α-helix content in HSA and HSA-PA to the same level. Dynamic light scattering (DLS) data revealed that both PA and EGCg increased HSA aggregation. EGCg increased HSA aggregation more significantly and promoted formation of aggregates that were more heterogenous. Any of these effects could impact the ability of serum albumin to transport and stabilize ligands including EGCg and other polyphenols.

Keywords: human serum albumin (HSA); fatty acid; polyphenol; protein conformation; protein aggregation; Förster resonance energy transfer (FRET)

---

1. Introduction

Human serum albumin (HSA) is the most abundant protein in human plasma with typical concentrations around 0.7 mM [1]. Serum albumin binds ligands such as fatty acids, glucose, drugs, excess copper, and hormones and aids in their transport, distribution, and metabolism. HSA comprises 585 amino acids, arranged mainly in alpha helices and stabilized by 17 disulfide bonds, leaving one free cysteine (Cys-34) [2,3]. The native protein has three structurally similar domains, and each domain can be divided into two subdomains, A and B. The two well-known drug-binding sites, Sudlow's sites I and II, are hydrophobic pockets in subdomains IIA and IIIA, respectively (Figure 1) [2,4]. The single Trp residue is in subdomain IIA. HSA in solution includes 60–70% monomers with the remainder forming dimers or larger oligomers [5]. Addition of ligands such as drugs and peptide linkers can promote protein dimerization and ultimately aggregation through cross-linking [6].

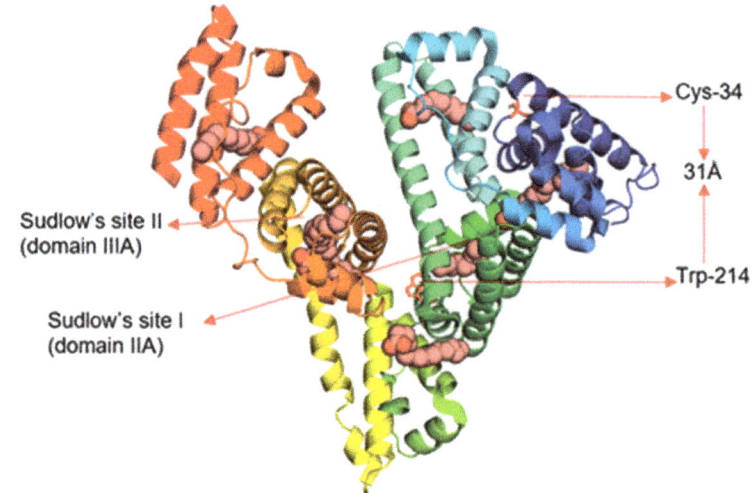

Figure 1. Human serum albumin (HSA) complexed with seven molecules of palmitic acid (PDB 1e7h). The different colors in HSA represent the six subdomains and the pink space filling models represent palmitic acid (PA). The Förster resonance energy transfer (FRET) distance between Trp-214 and Cys-34 in HSA is about 31 Å.

HSA is an allosteric protein that can undergo conformational changes in response to many factors including pH change and ligand binding [1,7,8]. For example, adding long chain saturated fatty acids such as octanoic acid or palmitic acid (PA) to HSA can introduce protein conformational changes [9,10]. Fatty acids bind to HSA in at least seven different binding sites involving all six subdomains of the protein, including Sudlow's sites I and II, with particularly high affinity for the latter (Figure 1) [9,10]. Allosteric changes induced by fatty acid binding alter the ability of HSA to transport other metabolites and drugs including Mn(III)-heme [11], the hormone thyroxine [12], insulin [13], and many sulfonylurea drugs [14].

Polyphenols are plant natural products comprising more than one aromatic ring and more than one phenolic functional group [15]. The characteristic phenolic reactivity is particularly potent for the high molecular weight polyphenols commonly called tannins. These compounds are powerful antioxidants [16], excellent metal chelators [17], and not only bind protein but also induce protein precipitation under favorable conditions [18]. It is widely reported that polyphenols such as epigallocatechin gallate (EGCg), the major polyphenolic compound found in green tea (Figure 2) [19], may have roles in preventing some chronic diseases when ingested as components of plant-based foods and beverages [20]. Although EGCg is a weak protein precipitating agent compared to higher molecular weight polyphenols, its potential to modify protein solubility could be related to its putative role in protein aggregation diseases such as Alzheimer's disease [21].

After ingestion, EGCg has limited stability in the gastrointestinal tract [22] but some of the material reaches the circulatory system where it is stabilized and carried by HSA [23]. An in vitro study has demonstrated that 1 mM EGCg is fully bound to 1.5 mM HSA within 5 min [23]. Fluorescence studies have shown that the binding constant (Kd) for EGCg and fatty acid-free HSA is about 14 μM [24,25]. Docking simulations and competitive binding experiments suggest that EGCg binds specifically to a hydrophobic pocket on HSA with more significant overlap of the nearby Sudlow's site I and less interaction with Sudlow's site II [26–30]. Recently, studies using isothermal titration calorimetry revealed that additional EGCg binds to the HSA surface non-specifically at multiple binding locations with very low affinity, with estimated Kd values in the millimolar range [27]. Upon binding, EGCg induces conformational changes in HSA and decreases its α-helical content [25,26,29,31]. Based on site-specific binding accompanied by activity changes, EGCg has been proposed as an allosteric effector for some proteins [32].

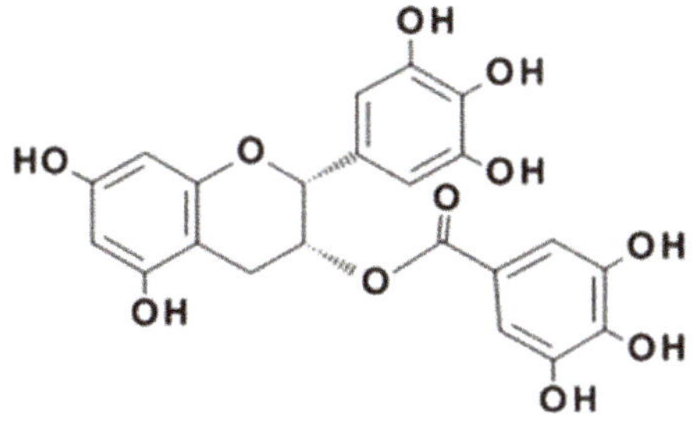

Figure 2. The chemical structure of the bioactive compound, (-)-epigallocatechin gallate (EGCg).

The one-on-one interaction between EGCg and serum albumin has been thoroughly examined [24–31]. However, under physiological conditions, serum albumin is typically loaded with one or more ligands consistent with its role as a carrier protein [1]. In vivo, EGCg would rarely encounter apo-serum albumin, but instead would bind the protein cooperatively or competitively with ligands such as fatty acids, metals, or drugs. To the best of our knowledge, nothing is known about how the endogenous ligands such as fatty acids or trace metals affect EGCg-serum albumin binding. A few studies have demonstrated that EGCg competes with 5-fluorouracil and related drugs for binding sites on serum albumin [30,33]. However, EGCg, resveratrol, and retinol simultaneously bind serum albumin at unique binding sites with minimal structural or chemical cross-talk between the ligands [34]. A knowledge gap that limits exploitation of the pharmaceutical potential of EGCg is poor understanding of how natural endogenous ligands of serum albumin combined with exogenous ligands such as EGCg affect the protein.

We hypothesized that EGCg may induce conformational changes not only in free HSA but also in the HSA-fatty acid complexes typically found in vivo. Furthermore, we proposed that EGCg would aggregate HSA-fatty acid complexes more efficiently than it aggregates the apoprotein, due to increased hydrophobicity of the HSA-fatty acid complex. We used several spectroscopic techniques including FRET (Förster resonance energy transfer), CD (circular dichroism), and DLS (dynamic light scattering) to study the details of the protein–polyphenol interaction in the presence and absence of the long chain saturated fatty acid PA. We suggest that this study will facilitate better understanding of the transport of the common dietary component and nutraceutical, EGCg, in human plasma as well as provide insights into potential interactions between EGCg and metabolites or drugs that are transported by serum albumin [11–14,30,35].

2. Materials and Methods

2.1. Materials

PA was purchased from Ultra Scientific (Philadelphia, PA 19122, USA), while fatty acid-free albumin from human serum was from Sigma-Aldrich (St. Louis, MO 63103, USA). EGCg (>95%) was obtained from Lipton Tea Co (Secacus, NJ 07094, USA). Reagent grade sodium phosphate was used to prepare 20 mM, pH 7.0 buffer used in all experiments, and guanidine hydrochloride was purchased from Sigma-Aldrich. HSA was labeled with one of two fluorescent probes, 6-propionyl-2-(dimethylamino) naphthalene (prodan) or 7-(diethyl amino)-4-methylcoumarin 3-maleimide (CPM), for FRET distance determination. Prodan was purchased from Life Technologies (Waltham, MA 02451, USA), while CPM was obtained from Chemodex (Buckingham, Bucks MK18 1EG, UK).

2.2. HSA Labeled with CPM

To label HSA with CPM, 10 μM of the protein was prepared in phosphate buffer and incubated with 6 M guanidine hydrochloride at 25 °C for 4 h [36]. CPM (150 μM) was prepared in 50% aqueous dimethyl sulfoxide and added to the solution to achieve 15 molar equivalents of HSA. The reaction was stopped after 4 h by adding 0.5 μL of 14 M β-mercaptoethanol [37]. The excess reagents and guanidine hydrochloride were removed by dialysis for 24 h at 4 °C against phosphate buffer, changing the dialysis buffer every 5 h. After dialysis, the samples were centrifuged at 17,000× g for 20 min to remove undissolved materials [37]. The Bradford assay [38] was used to determine the concentration of HSA and CPM-labeled HSA using bovine serum albumin as the standard protein. The incorporation of CPM was determined by direct spectrophotometry, using the extinction coefficient of 30,000 M^{-1} cm^{-1} at 383 nm [36]. The labeling efficiency was above 80%.

2.3. HSA Labeled with Prodan

Prodan was suspended in phosphate buffer and sonicated for an h to obtain a 300 μM solution [36]. The HSA-prodan complex was prepared by reacting 30 μM prodan with 1 μM HSA at 25 °C for 30 min [36]. The preparation had no fluorescence emission at 520 nm indicating there was no free prodan in the sample after the reaction was complete (Figure S1).

2.4. Reaction Mixtures

For FRET experiments, 1 μM HSA and HSA-CPM solutions were prepared with either PA (0 to 60 equivalents) or EGCg (0 to 25 equivalents). EGCg (0 to 25 equivalents) was also titrated into 1 μM HSA-PA and HSA-CPM-PA mixtures to assess its effect on the protein-fatty acid complex. Only HSA-prodan was titrated with different concentrations of EGCg; HSA-prodan-PA could not be used since PA replaces prodan on HSA [36]. Unlabeled HSA (5 μM) was prepared with EGCg (0 to 25 equivalents) and/or PA (0 to 60 equivalents) for CD and DLS measurements. Three independent sample replicates were prepared for FRET, CD, and DLS measurements.

PA has low solubility, which makes the preparation of HSA-PA complexes complicated. Curry's method was modified to obtain HSA-PA complexes [10]. After adding PA (2.5 mM) to 5 mL phosphate buffer, the mixture was incubated for 15 min at 50 °C. The PA suspension was then sonicated for 30 min before mixing with solutions of labeled or unlabeled HSA. After 30 min at room temperature the samples were centrifuged at 12,000× g for 10 min to remove undissolved PA.

EGCg was prepared in water and the concentration was determined by UV spectrometry based on its extinction coefficient (9700 M^{-1} cm^{-1} at 280 nm). EGCg-protein complexes were prepared by mixing the EGCg with the protein and incubating for 30 min at room temperature.

2.5. FRET Measurement

Fluorescence and absorbance measurements were used to calculate the FRET distance between Trp-214 and the prodan binding site or between Trp-214 and Cys-34-CPM. HSA and HSA-CPM or HSA and HSA-prodan was used as control for FRET distance comparison. A PerkinElmer LS 55 fluorescence spectrometer was used to measure sample fluorescence and an Agilent 8453 UV-visible spectrometer was used to measure sample absorbance at room temperature. The excitation wavelength for measuring fluorescence was 295 nm and both slit widths were set at 5 nm. The data were collected from 310 to 600 nm. Buffer blanks containing the appropriate concentration of CPM, PA, and/or EGCg were subtracted from the fluorescence data obtained with the reaction mixtures.

The distance between the intrinsic fluorophore Trp-214 (λ_{ex} = 295 nm, λ_{em} = 340 nm) and fluorescent labels within 60 Å can be determined using FRET. Free CPM has only weak fluorescence, but protein bound CPM is strongly fluorescent (λ_{ex} = 384 nm, λ_{em} = 470 nm). Free prodan has a strong emission at 520 nm that is easily differentiated from protein bound prodan (λ_{ex} = 360 nm, λ_{em} = 445 nm). In our experiments, excitation at 295 nm yielded the expected emission at 470 or 445 nm for

CPM- or prodan-labeled protein, respectively. To calculate the distance between prodan or CPM and the HSA intrinsic fluorophore Trp-214, the following Equations (1)–(4) were used:

$$E = \frac{R_0^6}{R_0^6 + R^6} \tag{1}$$

$$R_0 = 0.211(\kappa^2 \phi_d J \eta^{-4})^{1/6} \tag{2}$$

$$J(\lambda) = \int F_d(\lambda)\varepsilon_a(\lambda)\lambda^4 d\lambda / \int F_d(\lambda) d\lambda \tag{3}$$

$$E = 1 - F_{da}/F_d \tag{4}$$

According to Förster's theory [39–41], the efficiency of energy transfer (E) depends on the distance R (Å) between donor and acceptor fluorophores. R_0 is the distance between donor and acceptor when the energy transfer efficiency equals 50%. The refractive index of the medium, η, is 1.4 for this experiment. A value of 2/3 was used for κ^2, the orientation factor. The quantum yield of the donor in the absence of the acceptor, Φ_d, is 0.14 for Trp. In Equation (3), $F_d(\lambda)$ is the donor fluorescent intensity at the wavelength λ and $\varepsilon_a(\lambda)$ is the acceptor molar extinction coefficient [36,37]. J is the spectral overlap integral between the donor emission spectrum (Trp-214) and the acceptor absorbance spectrum (prodan or CPM), and was calculated by a|e—UV-Vis-IR spectral software FluorTools [42]. Equation (4) calculates the energy transfer value (E) based on the decrease of donor fluorescence intensity. F_d is the donor fluorescence intensity, measured in the absence of acceptor, while F_{da} is the donor fluorescence intensity when the acceptor is present.

2.6. CD Measurements

Circular dichroism was used to assess the secondary structure of HSA in the presence of EGCg and/or PA. A CD spectrometer MODEL 435 from AVIV biomedical, Inc. was used to measure the spectra at 25 °C in a quartz cuvette with 1 mm path length. The bandwidth was set at 1 nm, the speed was set at 50 nm min^{-1}, and the data were collected from 260 to 190 nm. Each spectrum was the average of three scans and three independent replicates of each sample were analyzed. For each sample, a background spectrum obtained with the appropriate concentration of EGCg was subtracted. To calculate the α-helical content in HSA at different conditions, the following Equations (5) and (6) were used [43].

$$MRE = \frac{\text{observed CD}}{C_p \times n \times l \times 10} \tag{5}$$

$$\alpha - \text{helix}(\%) = \frac{(-MRE_{208} - 4000)}{33000 - 4000} \times 100 \tag{6}$$

Using Equation (5), the mean residual ellipticity (MRE) can be calculated. C_p is the HSA molar concentration, n is the number of amino acid residues in HSA, and l mm is the path length. The α-helical content in protein was calculated using Equation (6). MRE_{208} is the calculated MRE value at 208 nm, 4000 is the MRE value of β-form and random coil conformations at 208 nm, and 33,000 is the MRE value of α-helices at 208 nm.

2.7. DLS Measurements

A Malvern dynamic light scattering system (DLS) was used to measure protein aggregation at 25 °C in a 10 mm disposable micro cuvette. All samples were equilibrated in the cuvette for 2 min, and then scanned 15 times. Three independent sample replicates were measured to check repeatability. The data output includes information about particle diameter, reported as size distribution, and relative quantity in solution, reported as intensity, for each peak.

3. Results

3.1. Interdomain Distances

FRET was used to determine the inter-domain distance between CPM at the single sulfhydryl group Cys-34 (subdomain IA) and Trp-214 (subdomain IIA) in apoprotein and in the presence of EGCg and/or PA. The fluorophore Trp is excited at 295 nm, and the emission of Trp at 340 nm can excite CPM that is within 60 Å of the Trp residue, with reduction in the Trp emission inversely proportional to the distance between the Trp and CPM (Δ_{em}, Figure 3a).

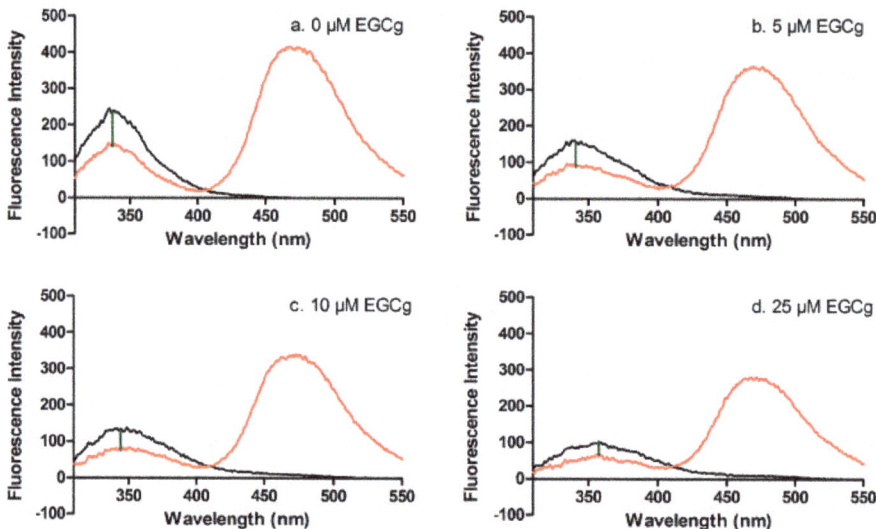

Figure 3. Fluorescence emission spectra for 1 µM HSA (black) and HSA-CPM (red) with 0 µM (**a**), 5 µM (**b**), 10 µM (**c**), 25 µM (**d**) EGCg. The difference in intensity at 340 nm (Δ_{em}) between HSA and HSA-CPM (green line) indicates the energy transfer.

EGCg complicated the FRET analysis because EGCg quenches the intrinsic fluorescence of HSA by binding to the protein near the Trp residue [26]. EGCg quenched HSA as expected (black lines, Figure 3a–d) and also quenched CPM-HSA (red lines, Figure 3a–d). In addition, EGCg induces a small red shift in the HSA or CPM-HSA emission spectrum (red lines, Figure 3a–d), probably because EGCg binds to HSA at the hydrophobic pocket and disrupts a salt bridge [44]. Since the EGCg-induced spectral changes applied to both HSA and HSA-CPM, they did not affect the calculated FRET energy transfer from Trp-214 to the CPM label. The data showed that EGCg diminished the Δ_{em}, indicating that the distance between Trp and CPM increased as more EGCg was added (Figure 3a–d). EGCg also quenched the Trp fluorescence of HSA-PA and HSA-CPM-PA, and decreased Δ_{em} in the PA-loaded protein (Figures S2–S4). The overlap of the HSA donor emission spectrum and the HSA-CPM acceptor absorption spectrum is suitable for J value calculation (Figure S5, Table S1).

In the apoprotein, the mean distance between Trp-214 and CPM-Cys-34 was 30.2 ± 1.5 Å, consistent with previous reports [36]. For each treatment, the change in distance between Trp-214 and CPM-Cys-34 was calculated by subtracting the distance determined in the presence of the ligand from the distance determined for apoHSA. A negative change in distance indicated that the ligand moved protein domains closer together, while a positive change in distance corresponded to increased distance between the labeled positions.

EGCg increased the distance between Trp-214 and CPM-Cys-34 in the absence of PA (Figure 4a, Table S2). The maximum increase detected was 2.8 ± 0.7 Å at a molar ratio of 25 EGCg to 1 HSA

(Figure 4a). The effect was dose dependent and fit a one binding site isotherm, with a calculated maximum change in distance of 3.2 Å and an apparent EC_{50} of 3.7 µM. The effect of PA on the protein was more complex (Figure 4b, Table S2). A small amount of PA added to the protein decreased the distance between Trp-214 and CPM-Cys-34 slightly, indicating that fatty acid at a low level made the structure somewhat more closed. Excess PA relaxed the protein, increasing the distance between the domains. The maximum decrease in distance induced by PA was 0.5 +/− 0.1 Å, while the maximum increase was 5.4 +/− 0.5 Å (Figure 4b). Palmitic acid induced larger changes in the protein structure than EGCg.

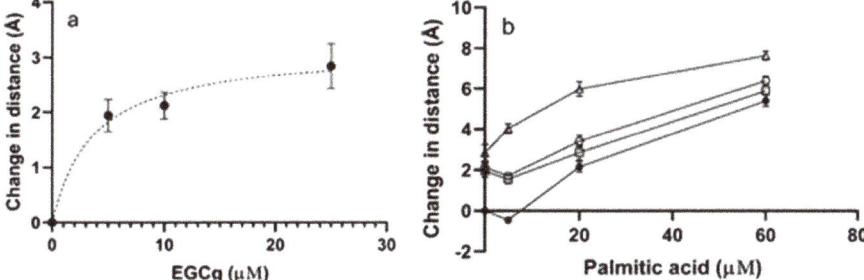

Figure 4. FRET distance between Trp-214 and CPM for HSA (1 µM with 0–25 µM of EGCg and/or 0–60 µM PA. The data points are means calculated from three replicates and the error bars indicate SEM. (**a**) FRET distance between Trp-214 and CPM with different concentrations of EGCg (0–25 µM). Bmax was 3.2 Å and apparent EC_{50} was 3.7 µM based on a saturation binding model for one ligand. (**b**) FRET distance between Trp-214 and CPM in the presence of PA (5, 20, 60 µM) and EGCg (0, 5, 10, 25 µM). ●, 0 µM EGCg; □, 5 µM EGCg; ○, 10 µM EGCg; Δ, 25 µM EGCg. The saturation binding model is not appropriate for PA binding to HSA.

When EGCg was added to the HSA-PA complexes, the distance between the domains generally increased but the magnitude of the increase was dependent on the amount of PA (Figure 4b, Table S2). The EGCg-induced increase in the Trp-to-Cys distance in HSA treated with 5 µM PA was less than the EGCg-induced increase for apoprotein, consistent with the ability of small amounts of PA to close the protein structure. The EGCg-induced increase in distance in HSA with excess PA was larger than the increase for apoprotein. The data suggested that the effects of the ligands were close to additive. For example, the distance increase was 7.6 +/− 0.4 Å for protein treated by the highest concentrations of PA and EGCg compared to 8.2 Å predicted for an additive effect.

FRET was used to determine the intra-domain flexibility of domain II by measuring energy transfer from Trp-214 to prodan, which binds to Sudlow's site I [36,45]. The method could only be used to examine the effects of EGCg because Sudlow's site I binds fatty acids, so addition of PA displaces the prodan label [36]. Although the EGCg binding site overlaps Sudlow's site I [26], addition of EGCg to prodan-labeled protein did not increase the fluorescence characteristic of free prodan (520 nm) (Figure S1), indicating that both ligands could simultaneously bind to the protein. The distance between prodan and Trp-214 in HSA was 25 ± 0.7 Å. Addition of EGCg did not change the distance (data not shown).

3.2. Secondary Structure

We used circular dichroism to assess the secondary structure of HSA in the presence of different ligands by monitoring the π-π* and n-π* transitions of peptide bonds in α-helices [46]. The CD spectra of HSA had negative bands at 209 and 222 nm (Figure 5 black line). When EGCg was added to HSA, the intensity of both bands decreased without any shift in the band maxima (Figure 5 red, green, and purple lines). The intensity change was larger in the band at 209 nm than that at 222 nm. EGCg decreased the α-helix content of HSA a small amount with a somewhat linear dose dependence (R^2 = 0.73) with about 7% less α-helix at the highest EGCg concentration compared to the apoprotein (Figure 6a, Table S3).

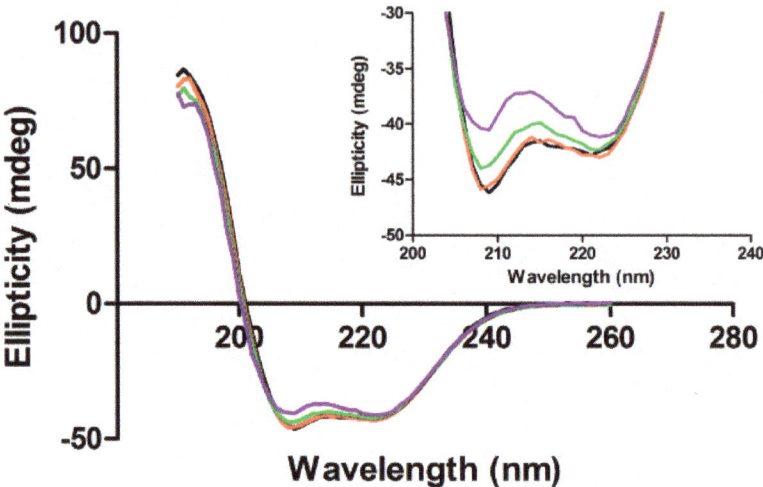

Figure 5. Circular dichroism (CD) spectra of HSA (5 µM) in the absence and presence of EGCg in 20 mM phosphate buffer at pH 7. The concentrations of EGCg were 0 µM (black), 25 µM (red), 50 µM (green), and 125 µM (purple).

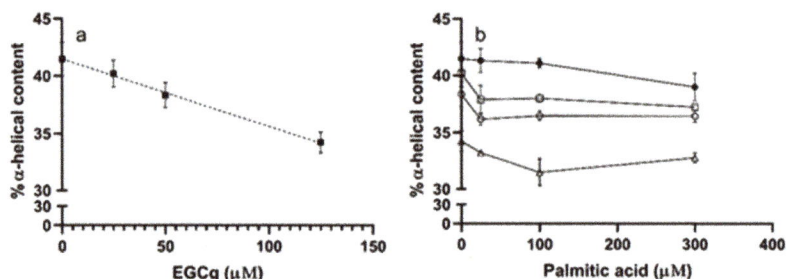

Figure 6. α-helical content in HSA (5 µM) with 0–125 µM of EGCg and/or 0–300 µM PA. Each point was calculated from three replicates with SEM for error bars. (**a**) HSA (5 µM) α-helical content with different concentrations of EGCg (0–125 µM). The line was linearly fit (Prism) to yield slope—0.29 and r^2 = 0.72. (**b**) HSA (5 µM) α-helical content in the presence of PA (0, 25, 100, 300 µM) and EGCg (0, 25, 50, 125 µM). ●, 0 µM EGCg; □, 25 µM EGCg; ○, 50 µM EGCg; ∆, 125 µM EGCg.

PA had little effect on the α-helix content of HSA (Figure 6b, Table S3), with the CD spectra in the presence of ligand similar to the spectrum of apoHSA (Figures S6–S8). The effect of EGCg on PA-treated protein is similar to the effect on apoprotein, with decreasing α-helix content proportional to the amount of EGCg added (Figure 6b, Table S3). Overall, EGCg induced a loss of the stability of α-helix, and affected protein secondary structure more significantly compared to PA.

3.3. Protein Size and Aggregation

Dynamic light scattering was used to monitor the changes in HSA size and aggregation upon interaction with EGCg and/or PA. Addition of PA did not change the size of the 5.3 nm HSA monomer

(Figure 7a), but addition of EGCg to HSA or PA-HSA moved the monomer peak slightly to the right (Figure 7b,c) indicating an approximately 0.4 nm increase in average diameter. However, the effects of EGCg were not very consistent. The conformational changes due to EGCg are small, and DLS measurements are dynamic, so it was reasonable that protein monomer size changes were not detected in every sample.

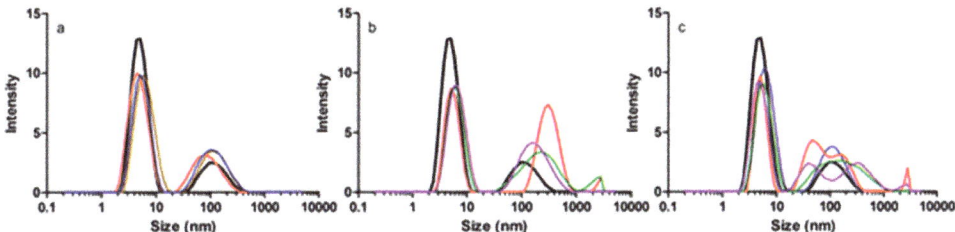

Figure 7. Dynamic light scattering (DLS) size distribution of (**a**) 5 μM HSA titrated with 0 μM (black), 25 μM (yellow), 100 μM (red), or 300 μM (blue) PA; (**b**) 5 μM HSA titrated with 0 μM (black), 25 μM (red), 50 μM (green), or 125 μM (purple) EGCg; (**c**) 5 μM HSA (black) and HSA (5 μM)-PA (300 μM) titrated with 0 μM (blue), 25 μM (red), 50 μM (green), or 125 μM (purple) EGCg.

DLS is more sensitive to protein aggregation and polydispersity than to small conformational changes. HSA solutions always contain some aggregated material, but EGCg or PA changed the area, position, and number of peaks of HSA aggregates. The effects of EGCg were not very reproducible within instrumental replicates or between chemical replicates, suggesting that the polyphenol induced heterogeneity. Generally, EGCg promoted aggregation more than PA. For apoHSA, about 23% of the total peak area was attributed to aggregates. PA increased the proportion of aggregates to about 35% (Figure 7a). EGCg increased the proportion of aggregates to about 45% for both apo HSA and PA-treated HSA (Figure 7b,c). Moreover, EGCg introduced more heterogeneity to HSA and HSA-PA based on the area, position, and number of different peaks representing aggregated protein (Figure 7b,c), while the aggregates induced by PA were more uniform (Figure 7a).

4. Discussion

At low concentrations the nonpolar polyphenol EGCg selectively binds a hydrophobic crevice in HSA with substantial overlap of Sudlow's site I [26–30]. However, EGCg also has polar characteristics (phenolic hydroxyl groups), and at sufficiently high concentration it can bind nonspecifically to the protein surface [27]. It has been reported that 25 equivalents of EGCg are required to fill all EGCg binding sites on HSA [27,31]. We used up to 25 equivalents of EGCg to study the maximal conformational changes that could be induced by EGCg to the apoprotein and the protein loaded with PA.

Our FRET results indicated that when EGCg binds the apoprotein, it increases the HSA inter-domain distance between the Trp-214 in subdomain IIA and CPM-Cys-34 in subdomain IA. EGCg did not change the organization of structural elements within domain II, monitored with prodan. The effect of EGCg was dose-dependent and saturable with a maximum change of 2.8 Å. The effect of EGCg on inter-domain distance was similar to but smaller than the effects of pH or denaturants on serum albumin [37,41]. The maximum increase of distance between domain I and domain II induced by increasing or decreasing pH is about 10 Å [37,41], while guanidine hydrochloride increases the inter-domain distance about 11 Å [37]. The data suggest that EGCg binding does not fully denature the protein, but disrupts interdomain interactions by altering the polar environment near the hydrophobic pocket and Sudlow's site I, thought to be a conformationally adaptable region of the protein [47,48]. The saturable nature of the EGCg effect supports the idea that this ligand modulates inter-domain distances by specific binding at the hydrophobic pocket and Sudlow's site I. Even though EGCg at high concentration could nonspecifically bind to HSA surface, such nonspecific binding may not significantly affect protein conformation.

The effects of PA on inter-domain distances are more complex, with low levels of PA decreasing the distance between subdomains IA and IIA, and saturating levels of PA increasing the distance. In our experiment, up to 60 equivalents of PA were added to HSA to ensure that all sites are occupied [9,36]. The variable allosteric effects of PA on protein structure are a consequence of the protein's ability to bind fatty acid at several widely distributed sites on the protein [9,10], such as FA4 and FA5 in domain III [49,50] and FA2 in subdomain IA [51], with different effects on structure as more sites are filled. Previous studies have confirmed a dose dependent effect of fatty acids on serum albumin structure, with the structure closing at low PA concentrations due to rotation of the subdomain IB-IIA inter-domain helix [52]. At higher PA concentrations the subdomain IIB-IIIA inter-domain helix tilts slightly, resulting in rotation of domains I and III away from domain II and opening up the central crevice about 10 Å [9,52,53]. These changes are consistent with our FRET data that shows that 60 equivalents of PA increases HSA inter-domain distance between domain IA and IIA up to 5.4 Å.

When added to the PA-serum albumin complex, EGCg opens the protein structure more than when the fatty acid is not present. It has been reported that Sudlow's site I ligand binding is enhanced in the protein-fatty acid complex [11]. The increased interdomain distance induced by EGCg may indicate that EGCg more efficiently binds PA-serum albumin compared to apoprotein.

Despite the ability of PA to alter inter-domain protein distances in HSA, the CD results suggested that PA had little effect on α-helix content of the protein. Our data are consistent with other studies that have established that the overall high α-helical character of serum albumin is slightly decreased only when very high levels of fatty acids are added to the protein [36,54]. The minor changes in helical content could be due to high levels of hydrophobic ligands [55] or could be due to protein unfolding induced by micelles [56].

Adding EGCg to HSA decreased the α-helical content in a dose-dependent fashion consistent with previous studies [24,25]. Unlike its effect on inter-domain distance, EGCg affected α-helix content in a non-saturable fashion. This suggests that the effects on helical content are due to nonspecific binding, probably at the protein surface through hydrogen bonding between phenolic hydroxyls and peptide carbonyls [27,57]. The decrease in α-helical content was similar to that observed with pH changes that disrupt hydrogen bonding and electrostatic interactions in the protein [41].

When added to the HSA-PA complex, the effects of EGCg on the protein α-helical content were similar to its effect on apoHSA. EGCg overwhelmed the small effects of PA on the protein secondary structure. PA binds at specific interior sites on the protein by hydrophobic and electrostatic interactions [9,58], while EGCg binds to a specific interior hydrophobic pocket, and nonspecifically to the surface at least in part via hydrogen bonds at the peptide backbone [26,27]. Therefore, polyphenols such as EGCg are more likely to disrupt α-helices.

The ability of EGCg to precipitate protein is more well-known than its effect on protein secondary and tertiary structure. EGCg is a less effective precipitating agent than higher molecular weight polyphenols (tannins), requiring a stoichiometric ratio of 125:1 to achieve sufficient protein coating and crosslinking to initiate precipitation [28]. Our DLS data revealed that at lower EGCg:protein ratios (5:1, 10:1, 25:1), the polyphenol stimulates HSA aggregation without precipitation. The formation of aggregates was not strongly concentration dependent and yielded highly polydisperse mixtures. Unlike the site-specific binding event by EGCg that results in changes in domain substructures in the protein, aggregation apparently takes place when excess EGCg binds nonspecifically to the surface of the protein. Binding to the surface decreases helical content and starts to denature the protein, and may also create "hot spots" of hydrophobicity or masked surface charge that promote aggregate formation [59]. Since surface binding is nonspecific, some binding events may not significantly change the protein, resulting in polydisperse mixtures that form precipitates only when the polyphenol is in significant excess over the protein so that many nonspecific surface sites are filled and cross-links can form between proteins.

PA also increased protein aggregation, consistent with previous studies [60]. Compared to EGCg, PA induced formation of more homogeneous aggregates. PA binds to HSA at specific sites on the interior of the protein, mainly through hydrophobic interactions. Loading the protein with fatty acid could cause a systematic change in hydrophobicity that could increase formation of aggregates. Ligands such as fatty acids that have site-specific binding may form aggregates with defined stoichiometries, leading to mixtures that are less polydisperse than the populations of aggregates generated by a nonspecific ligand such as EGCg.

EGCg aggregated HSA-PA complexes in a fashion similar to its behavior with apoHSA. The result suggests that binding PA at interior sites of the protein does not affect EGCg binding at the surface of the protein. EGCg is well-suited to interact with hydrophobic and hydrophilic sites on the protein, because of its dual nature as an aromatic, hydrophobic compound with numerous polar hydroxyl functional groups. Thus, it is able to aggregate either the apoprotein or the less polar PA-loaded protein, in each case forming hetero-disperse aggregates due to nonspecific surface binding on the protein. Although it has been suggested that EGCg has therapeutic potential because of its ability to remodel fibrillar proteins [61,62], our data indicate a rather nonselective effect, suggesting that the potential for EGCg to control protein fibril formation in disease states may be limited.

5. Conclusions

This study shows that the dietary polyphenol, EGCg, and fatty acids, such as PA, simultaneously bind to serum albumin and influence the structure of the protein. The highly hydrophobic fatty acid and the amphipathic polyphenol appear to have unique binding characteristics with the protein, leading to the conclusion that in vivo the fatty acid-loaded serum albumin probably does transport EGCg. However, the additive effect of the two ligands on protein features such as the interdomain distance between subdomain IA and subdomain IIA could have important consequences for drugs whose binding at Sudlow's site I is allosterically controlled by fatty acids. Our study demonstrated that EGCg influences protein secondary structure while the effects of PA on HSA are related to protein conformation. Further studies of drug binding, transport, and release by HSA in the presence of both EGCg and fatty acids are critical to a better understanding of EGCg, which is not only a natural dietary component but also a popular dietary supplement.

Supplementary Materials: The following are available online at http://www.mdpi.com/2218-273X/9/11/705/s1, Figure S1: Fluorescence spectra of HSA and HSA-prodan with different concentrations of EGCg. Figure S2: Fluorescence emission spectra for HSA (1 µM)-PA (5 µM) and HSA-CPM (1 µM)-PA (5 µM) with 0 µM (a), 5 µM (b), 10 µM (c), 25 µM (d) EGCg. Figure S3: Fluorescence emission spectra for HSA (1 µM)-PA (20 µM) and HSA-CPM (1 µM)-PA (20 µM) with 0 µM (a), 5 µM (b), 10 µM (c), 25 µM (d) EGCg. Figure S4: Fluorescence emission spectra for HSA (1 µM)-PA (60 µM) and HSA-CPM (1 µM)-PA (60 µM) with 0 µM (a), 5 µM (b), 10 µM (c), 25 µM (d) EGCg. Figure S5: Overlap of the donor emission spectrum (Trp-214) and the acceptor absorption spectrum (HAS-CPM). Figure S6: CD spectra of 5 µM HSA and 5 µM HSA in the presence of 25 µM PA, and 0–125 µM EGCg. Figure S7: CD spectra of 5 µM HSA and 5 µM HSA in the presence of 100 µM PA, and 0–125 µM EGCg. Figure S8: CD spectra of 5 µM HSA and 5 µM HSA in the presence of 300 µM PA, and 0–125 µM EGCg. Table S1: J, R_0, and R values for HSA, HSA-PA, and HSA-PA-EGCg for three replicates. Table S2: The change in distance (Å) between Trp-214 and CPM induced by addition of EGCg and/or palmitic acid. Table S3: The α-helical content (%) of HSA with various amounts of EGCg and/or palmitic acid.

Author Contributions: Conceptualization, A.E.H.; Data curation, X.S.; Formal analysis, X.S.; Funding acquisition, A.E.H.; Investigation, X.S. and H.N.F.; Methodology, X.S.; Project administration, A.E.H.; Supervision, A.E.H.; Validation, X.S.; Writing—original draft, X.S.; Writing—review and editing, A.E.H.

Funding: This research was funded by the National Science Foundation, grant number 1750189. The APC was funded by Miami University.

Acknowledgments: We acknowledge C. Scott Hartley for the use of the fluorimeter, Gary A. Lorigan for use of the DLS instrument, and Indra Sahu for assistance with CD and DLS.

Conflicts of Interest: The authors declare no conflict of interest.

References

1. Fanali, G.; di Masi, A.; Trezza, V.; Marino, M.; Fasano, M.; Ascenzi, P. Human serum albumin: From bench to bedside. *Mol. Asp. Med.* **2012**, *33*, 209–290. [CrossRef] [PubMed]
2. He, X.M.; Carter, D.C. Atomic structure and chemistry of human serum albumin. *Nature* **1992**, *358*, 209–215. [CrossRef] [PubMed]
3. Sugio, S.; Kashima, A.; Mochizuki, S.; Noda, M.; Kobayashi, K. Crystal structure of human serum albumin at 2.5 A resolution. *Protein Eng. Des. Sel.* **1999**, *12*, 439–446. [CrossRef] [PubMed]
4. Sudlow, G.; Birkett, D.J.; Wade, D.N. Characterization of 2 specific drug binding sites on human serum albumin. *Mol. Pharmacol.* **1975**, *11*, 824–832. [PubMed]
5. Nemashkalova, E.L.; Permyakov, E.A.; Permyakov, S.E.; Litus, E.A. Modulation of linoleic acid-binding properties of human serum albumin by divalent metal cations. *BioMetals* **2017**, *30*, 341–353. [CrossRef]
6. Taguchi, K.; Victor, T.G.C.; Maruyama, T.; Otagiri, M. Pharmaceutical aspects of the recombinant human serum albumin dimer: Structural characteristics, biological properties, and medical applications. *J. Pharm. Sci.* **2012**, *101*, 3033–3046. [CrossRef]
7. Soltys, B.J.; Hsia, J.C. Human serum albumin.1. Relationship of fatty-acid and bilirubin binding-sites and nature of fatty-acid allosteric effects–monoanionic spin label study. *J. Biol. Chem.* **1978**, *253*, 3023–3028.
8. Kun, R.; Szekeres, M.; Dekany, I. Isothermal titration calorimetric studies of the pH induced conformational changes of bovine serum albumin. *J. Therm. Anal. Calorim.* **2009**, *96*, 1009–1017. [CrossRef]
9. Bhattacharya, A.A.; Grune, T.; Curry, S. Crystallographic analysis reveals common modes of binding of medium and long-chain fatty acids to human serum albumin. *J. Mol. Biol.* **2000**, *303*, 721–732. [CrossRef]
10. Curry, S.; Mandelkow, H.; Brick, P.; Franks, N. Crystal structure of human serum albumin complexed with fatty acid reveals an asymmetric distribution of binding sites. *Nat. Struct. Biol.* **1998**, *5*, 827–835. [CrossRef]
11. Fanali, G.; Fesce, R.; Agrati, C.; Ascenzi, P.; Fasano, M. Allosteric modulation of myristate and Mn(III)heme binding to human serum albumin. *FEBS J.* **2005**, *272*, 4672–4683. [CrossRef] [PubMed]
12. Fasano, M.; Curry, S.; Terreno, E.; Galliano, M.; Fanali, G.; Narciso, P.; Notari, S.; Ascenzi, P. The extraordinary ligand binding properties of human serum albumin. *IUBMB Life* **2005**, *57*, 787–796. [CrossRef] [PubMed]
13. Bomba-Opon, D.; Wielgos, M.; Szymanska, M.; Bablok, L. Effects of free fatty acids on the course of gestational diabetes mellitus. *Neuroendocrinol. Lett.* **2006**, *27*, 277–280. [PubMed]
14. Anguizola, J.; Matsuda, R.; Barnaby, O.S.; Hoy, K.S.; Wa, C.L.; DeBolt, E.; Koke, M.; Hage, D.S. Glycation of human serum albumin. *Clin. Chim. Acta* **2013**, *425*, 64–76. [CrossRef]
15. Quideau, S.; Deffieux, D.; Douat-Casassus, C.; Pouysegu, L. Plant polyphenols: Chemical properties, biological activities, and synthesis. *Angew. Chem. Int. Ed.* **2011**, *50*, 586–621. [CrossRef]
16. Hagerman, A.E.; Riedl, K.M.; Jones, G.A.; Sovik, K.N.; Ritchard, N.T.; Hartzfeld, P.W.; Riechel, T.L. High molecular weight plant polyphenolics (tannins) as biological antioxidants. *J. Agric. Food Chem.* **1998**, *46*, 1887–1892. [CrossRef]
17. Zhang, L.; Liu, R.; Gung, B.W.; Tindall, S.; Gonzalez, J.M.; Halvorson, J.J.; Hagerman, A.E. Polyphenol-aluminum complex formation: Implications for aluminum tolerance in plants. *J. Agric. Food Chem.* **2016**, *64*, 3025–3033. [CrossRef]
18. Hagerman, A.E. Fifty years of polyphenol-protein complexes. *Rec. Adv. Polyphen. Res.* **2012**, *3*, 71–97.
19. Higdon, J.V.; Frei, B. Tea catechins and polyphenols: Health effects, metabolism, and antioxidant functions. *Crit. Rev. Food Sci. Nutr.* **2003**, *43*, 89–143. [CrossRef]
20. Del Rio, D.; Rodriguez-Mateos, A.; Spencer, J.P.E.; Tognolini, M.; Borges, G.; Crozier, A. Dietary (poly)phenolics in human health: Structures, bioavailability, and evidence of protective effects against chronic diseases. *Antioxid. Redox Signal.* **2013**, *18*, 1818–1892. [CrossRef]
21. Weinreb, O.; Amit, T.; Mandel, S.; Youdim, M.B.H. Neuroprotective molecular mechanisms of (-)-epigallocatechin-3-gallate: A reflective outcome of its antioxidant, iron chelating and neuritogenic properties. *Genes Nutr.* **2009**, *4*, 283–296. [CrossRef] [PubMed]
22. Krook, M.A.; Hagerman, A.E. Stability of polyphenols epigallocatechin gallate and pentagalloyl glucose in a simulated digestive system. *Food Res. Int.* **2012**, *49*, 112–116. [CrossRef] [PubMed]
23. Zinellu, A.; Sotgia, S.; Scanu, B.; Forteschi, M.; Giordo, R.; Cossu, A.; Posadino, A.M.; Carru, C.; Pintus, G. Human serum albumin increases the stability of green tea catechins in aqueous physiological conditions. *PLoS ONE* **2015**, *10*, e0134690. [CrossRef] [PubMed]

24. Trnkova, L.; Bousova, I.; Stankova, V.; Drsata, J. Study on the interaction of catechins with human serum albumin using spectroscopic and electrophoretic techniques. *J. Mol. Struct.* **2011**, *985*, 243–250. [CrossRef]
25. Maiti, T.K.; Ghosh, K.S.; Dasgupta, S. Interaction of (-)-epigallocatechin-3-gallate with human serum albumin: Fluorescence, Fourier transform infrared, circular dichroism, and docking studies. *Proteins* **2006**, *64*, 355–362. [CrossRef]
26. Li, M.; Hagerman, A.E. Role of the flavan-3-ol and galloyl moieties in the interaction of (-)-epigallocatechin gallate with serum albumin. *J. Agric. Food Chem.* **2014**, *62*, 3768–3775. [CrossRef]
27. Eaton, J.D.; Williamson, M.P. Multi-site binding of epigallocatechin gallate to human serum albumin measured by NMR and isothermal titration calorimetry. *Biosci. Rep.* **2017**, *37*. [CrossRef]
28. Li, M.; Hagerman, A.E. Interactions Between Plasma Proteins and Naturally Occurring Polyphenols. *Curr. Drug Metab.* **2013**, *14*, 432–445. [CrossRef]
29. Nozaki, A.; Hori, M.; Kimura, T.; Ito, H.; Hatano, T. Interaction of polyphenols with proteins: Binding of (-)-epigallocatechin gallate to serum albumin, estimated by induced circular dichroism. *Chem. Pharm. Bull. (Tokyo)* **2009**, *57*, 224–228. [CrossRef]
30. Yuan, L.X.; Liu, M.; Liu, G.Q.; Li, D.C.; Wang, Z.P.; Wang, B.Q.; Han, J.; Zhang, M. Competitive binding of (-)-epigallocatechin-3-gallate and 5-fluorouracil to human serum albumin: A fluorescence and circular dichroism study. *Spectrochim. Acta A* **2017**, *173*, 584–592. [CrossRef]
31. Save, S.N.; Choudhary, S. Elucidation of energetics and mode of recognition of green tea polyphenols by human serum albumin. *J. Mol. Liq.* **2018**, *265*, 807–817. [CrossRef]
32. Li, M.; Li, C.H.; Allen, A.; Stanley, C.A.; Smith, T.J. The structure and allosteric regulation of mammalian glutamate dehydrogenase. *Arch. Biochem. Biophys.* **2012**, *519*, 69–80. [CrossRef] [PubMed]
33. Yuan, L.X.; Liu, M.; Shi, Y.B.; Yan, H.; Han, J.; Liu, L.Y. Effect of (-)-epicatechin-3-gallate and (-)-epigallocatechin-3-gallate on the binding of tegafur to human serum albumin as determined by spectroscopy, isothermal titration calorimetry, and molecular docking. *J. Biomol. Struct. Dyn.* **2019**, *37*, 2776–2788. [CrossRef] [PubMed]
34. Wu, Y.; Cheng, H.; Chen, Y.T.; Chen, L.Y.; Fang, Z.; Liang, L. Formation of a Multiligand Complex of Bovine Serum Albumin with Retinol, Resveratrol, and (-)-Epigallocatechin-3-gallate for the Protection of Bioactive Components. *J. Agric. Food Chem.* **2017**, *65*, 3019–3030. [CrossRef] [PubMed]
35. Ascenzi, P.; Bocedi, A.; Notari, S.; Menegatti, E.; Fasano, M. Heme impairs allosterically drug binding to human serum albumin Sudlow's site I. *Biochem. Biophys. Res. Commun.* **2005**, *334*, 481–486. [CrossRef]
36. Krishnakumar, S.S.; Panda, D. Spatial relationship between the prodan site, Trp-214, and Cys-34 residues in human serum albumin and loss of structure through incremental unfolding. *Biochemistry* **2002**, *41*, 7443–7452. [CrossRef]
37. Chowdhury, R.; Chattoraj, S.; Sen Mojumdar, S.; Bhattacharyya, K. FRET between a donor and an acceptor covalently bound to human serum albumin in native and non-native states. *Phys. Chem. Chem. Phys.* **2013**, *15*, 16286–16293. [CrossRef]
38. Bradford, M.M. A rapid and sensitive method for the quantitation of microgram quantitites of protein utilizing the principle of protein-dye binding. *Anal. Biochem.* **1976**, *72*, 248–254. [CrossRef]
39. Wu, P.G.; Brand, L. Resonance energy-transfer–methods and applications. *Anal. Biochem.* **1994**, *218*, 1–13. [CrossRef]
40. Stryer, L. Fluorescence energy-transfer as a spectroscopic ruler. *Annu. Rev. Biochem.* **1978**, *47*, 819–846. [CrossRef]
41. Shaw, A.K.; Pal, S.K. Spectroscopic studies on the effect of temperature on pH-induced folded states of human serum albumin. *J. Photochem. Photobiol. B Biol.* **2008**, *90*, 69–77. [CrossRef] [PubMed]
42. Preus, S.; Kilsa, K.; Miannay, F.A.; Albinsson, B.; Wilhelmsson, L.M. FRETmatrix: A general methodology for the simulation and analysis of FRET in nucleic acids. *Nucleic Acids Res.* **2013**, *41*, e18. [CrossRef] [PubMed]
43. Chen, Y.H.; Yang, J.T.; Martinez, H.M. Determination of secondary structures of proteins by circular dichroism and optical rotatory dispersion. *Biochemistry* **1972**, *11*, 4120–4131. [CrossRef] [PubMed]
44. Vivian, J.T.; Callis, P.R. Mechanisms of tryptophan fluorescence shifts in proteins. *Biophys. J.* **2001**, *80*, 2093–2109. [CrossRef]
45. Moreno, F.; Cortijo, M.; Gonzalez-Jimenez, J. The fluorescent probe prodan characterizes the warfarin binding site on human serum albumin. *Photochem. Photobiol.* **1999**, *69*, 8–15. [CrossRef] [PubMed]

46. Kandagal, P.B.; Ashoka, S.; Seetharamappa, J.; Shaikh, S.M.T.; Jadegoud, Y.; Ijare, O.B. Study of the interaction of an anticancer drug with human and bovine serum albumin: Spectroscopic approach. *J. Pharm. Biomed. Anal.* **2006**, *41*, 393–399. [CrossRef]
47. Yamasaki, K.; Maruyama, T.; KraghHansen, U.; Otagiri, M. Characterization of site I on human serum albumin: Concept about the structure of a drug binding site. *Biochim. Biophys. Acta* **1996**, *1295*, 147–157. [CrossRef]
48. Fehske, K.J.; Schlafer, U.; Wollert, U.; Muller, W.E. Characterization of an important drug-binding area on human serum albumin including the high-affinity binding-sites of warfarin and azapropazone. *Mol. Pharmacol.* **1982**, *21*, 387–393.
49. Chuang, V.T.G.; Otagiri, M. How do fatty acids cause allosteric binding of drugs to human serum albumin? *Pharm. Res.* **2002**, *19*, 1458–1464. [CrossRef]
50. Rietbrock, N.; Menke, G.; Reuter, G.; Lassmann, A.; Schmeidl, R. Influence of palmitate and oleate on the binding of warfarin to human serum albumin–stopped-flow studies. *J. Clin. Chem. Clin. Biol.* **1985**, *23*, 719–723. [CrossRef]
51. Petitpas, I.; Bhattacharya, A.A.; Twine, S.; East, M.; Curry, S. Crystal structure analysis of warfarin binding to human serum albumin–Anatomy of Drug Site I. *J. Biol. Chem.* **2001**, *276*, 22804–22809. [CrossRef] [PubMed]
52. Ascenzi, P.; Fasano, M. Allostery in a monomeric protein: The case of human serum albumin. *Biophys. Chem.* **2010**, *148*, 16–22. [CrossRef] [PubMed]
53. Curry, S.; Brick, P.; Franks, N.P. Fatty acid binding to human serum albumin: New insights from crystallographic studies. *BBA-Mol. Cell. Biol. Lipids* **1999**, *1441*, 131–140. [CrossRef]
54. Oleszko, A.; Hartwich, J.; Gasior-Glogowska, M.; Olsztynska-Janus, S. Changes of albumin secondary structure after palmitic acid binding. FT-IR spectroscopic study. *Acta Bioeng. Biomech.* **2018**, *20*, 59–64.
55. Ghosh, S.; Dey, J. Binding of fatty acid amide amphiphiles to bovine serum albumin: Role of amide hydrogen bonding. *J. Phys. Chem. B* **2015**, *119*, 7804–7815. [CrossRef]
56. Otzen, D.E. Protein unfolding in detergents: Effect of micelle structure, ionic strength, pH, and temperature. *Biophys. J.* **2002**, *83*, 2219–2230. [CrossRef]
57. Hagerman, A.E.; Butler, L.G. The specificity of tannin protein interactions. *J. Biol. Chem.* **1981**, *256*, 4494–4497.
58. Spector, A.A. Fatty-acid binding to plasma albumin. *J. Lipid Res.* **1975**, *16*, 165–179.
59. Roberts, C.J. Therapeutic protein aggregation: Mechanisms, design, and control. *Trends Biotechnol.* **2014**, *32*, 372–380. [CrossRef]
60. White, J.; Hess, D.; Raynes, J.; Laux, V.; Haertlein, M.; Forsyth, T.; Jeyasingham, A. The aggregation of "native" human serum albumin. *Eur. Biophys. J. Biophys.* **2015**, *44*, 367–371. [CrossRef]
61. Chandrashekaran, I.R.; Adda, C.G.; MacRaild, C.A.; Anders, R.F.; Norton, R.S. EGCG disaggregates amyloid-like fibrils formed by Plasmodium falciparum merozoite surface protein 2. *Arch. Biochem. Biophys.* **2011**, *513*, 153–157. [CrossRef] [PubMed]
62. Stenvang, M.; Dueholm, M.S.; Vad, B.S.; Seviour, T.; Zeng, G.H.; Geifman-Shochat, S.; Sondergaard, M.T.; Christiansen, G.; Meyer, R.L.; Kjelleberg, S.; et al. Epigallocatechin gallate remodels overexpressed functional amyloids in pseudomonas aeruginosa and increases biofilm susceptibility to antibiotic treatment. *J. Biol. Chem.* **2016**, *291*, 26540–26553. [CrossRef] [PubMed]

© 2019 by the authors. Licensee MDPI, Basel, Switzerland. This article is an open access article distributed under the terms and conditions of the Creative Commons Attribution (CC BY) license (http://creativecommons.org/licenses/by/4.0/).

Article

Optimization of Catechin and Proanthocyanidin Recovery from Grape Seeds Using Microwave-Assisted Extraction

Jing Chen [1], W. P. D. Wass Thilakarathna [2], Tessema Astatkie [3] and H. P. Vasantha Rupasinghe [2,*]

[1] Institute of TCM and Natural Products, Key Laboratory of Combinatorial Biosynthesis and Drug Discovery (Ministry of Education), School of Pharmaceutical Sciences, Wuhan University, 185 East Lake Road, Wuhan 430071, China; chenjingz@whu.edu.cn
[2] Department of Plant, Food, and Environmental Sciences, Faculty of Agriculture, Dalhousie University, Truro, NS B2N 5E3, Canada; ws714884@dal.ca
[3] Faculty of Agriculture, Dalhousie University, Truro, NS B2N 5E3, Canada; astatkie@dal.ca
* Correspondence: vrupasinghe@dal.ca; Tel.: +1-902-893-6623

Received: 3 January 2020; Accepted: 2 February 2020; Published: 5 February 2020

Abstract: Grape seed extract (GSE) is a rich source of condensed flavonoid tannins, also called proanthocyanidins (PACs). The high molecular weight of polymeric PAC limits their biological activity due to poor bioavailability. The present study was undertaken to explore the potential applicability of microwave-assisted extraction (MAE) to convert GSE-PAC into monomeric catechins. A central composite design (CCD) was used to optimize the processing conditions for the MAE. The maximum total yield of monomeric catechins (catechin, epicatechin, and epicatechin gallate) and PAC were 8.2 mg/g dry weight (DW) and 56.4 mg catechin equivalence (CE)/g DW, respectively. The optimized MAE condition was 94% ethanol, 170 °C temperature, and a duration of 55 min. Compared to the results for PACs extracted via conventional extraction (Con) (94% ethanol; shaking at 25 °C for 55 min), MAE yielded 3.9-fold more monomeric catechins and 5.5-fold more PACs. The MAE showed higher antioxidant capacity and α-glucosidase inhibitory activity than that of the conventional extract, suggesting the potential use of the MAE products of grape seeds as a functional food ingredient and nutraceutical.

Keywords: flavanols; condensed tannin; *Vitis vinifera*; microwave-extraction; antioxidant; glucosidase; cytotoxicity

---

1. Introduction

Catechins, also called flavanols or flavan-3-ols, are a sub-group of flavonoids that are found in some plant-based foods [1]. The commonly found catechins include catechin, epicatechin (EC), epicatechin gallate (ECG), epigallocatechin (EGC), and epigallocatechin gallate (EGCG). Over the past two decades, catechins have attracted interest due to their biological activities, such as antioxidant, antibacterial, and anti-inflammatory properties, among many others [2,3]. Moreover, increasing number of studies have demonstrated that catechins also possess anticarcinogenic activity in many experimental systems and many kinds of organs, including the intestine, lung, liver, pancreas, skin, prostate, breast, and cervix [4–9]. The cancer-preventive activity of catechins has been suggested to be due to their antioxidant activity and the modulation of multiple cellular signaling pathways [10,11].

Grapes (*Vitis vinifera* L.) are one of the most abundant fruit crops in the world, with an annual production of 69 million metric tons [12]. Grape seeds comprise 5% by mass of grapes and are the major industrial byproducts of grape-processing industries such as grape juice and wine. Grape seeds are a good source of phenolic compounds such as gallic acid, catechin, epicatechin, and proanthocyanidins (PACs, condensed tannins) [12]. Grape seed extracts (GSE) are widely consumed as a dietary supplement based on their potent antioxidant, anticancer, antimicrobial, anti-aging, and anti-inflammatory activities, and are generally recognized as safe by the US Food and Drug Administration (US FDA) [13]. However, PACs with a degree of polymerization over four are not absorbable because of their large molecular size. It has been reported that GSE contains a heterogeneous mixture of PAC monomers (5–30%), oligomers (17–63%), and polymers (11–39%) [14].

In the present study, microwave energy was explored as a tool to enhance the extraction of PAC as well as to increase the generation of monomers through depolymerization of PAC. Microwave-assisted extraction (MAE) is a fast and efficient bioactive extraction method that is based on the direct impact on polar compounds [15,16]. Electromagnetic energy, in the frequency range of 300 MHz to 300 GHz, is transferred to heat following ionic conduction and dipole rotation. MAE has been used to enhance the extraction of active compounds from many plant matrixes, including grape seeds [17–19]. MAE heats the matrix internally and externally without a thermal gradient; therefore, biomolecules can be extracted efficiently and protectively [18]. MAE is advantageous over conventional extraction techniques, with improved efficiency, reduced extraction time, rapid and volumetric heating of the absorbing medium, low solvent consumption, higher selectivity of target molecules, and a high potential for automation [19]. However, the effects of MAE conditions such as solvents and temperature on the generation of monomeric catechins have not been reported. Therefore, this study aimed to (1) optimize MAE conditions for the recovery of PACs with maximum monomeric catechins from grape seeds, and (2) assess antioxidant capacity, α-glucosidase inhibitory activity, and cytotoxicity of the MAE products.

2. Materials and Methods

2.1. Materials and Chemicals

Grape seed powder used in this study was provided by Royal Grapeseed, Milton, NY, USA. The grape seeds were from a mixture of commercial grape varieties of *Vitis vinifera*, *V. labrusca*, and hybrids of native American species with *V. vinifera*. The liquid chromatography standards used for the study were obtained as follows: (−)-epicatechin, (+)-catechin, epigallocatechin (EGC), epicatechin gallate (ECG), epigallocatechin gallate (EGCG), and procyanidin B1 and B2 were from ChromaDex (Santa Ana, CA, USA). High-performance liquid chromatography (HPLC) grade methanol, acetonitrile, and formic acid as well as 2-mercaptoethanol, acetic acid, citric acid, gallic acid, hydrochloric acid, sodium sulfite, and methylcellulose were obtained from Sigma-Aldrich (Mississauga, ON, Canada).

For the cell culture experiments, human fetal hepatic (WRL-68; ATCC® CL-48™) and human hepatocellular carcinoma (HepG2; ATCC® HB-8065™) cells were purchased from American Type Culture Collection (Manassas, VA, USA). Minimum essential medium eagle (MEME), fetal bovine serum (FBS), penicillin-streptomycin, L-glutamine, dimethyl sulfoxide (DMSO), phosphate-buffered saline (PBS), trypan blue stain, and phenazine methosulphate (PMS) were purchased from Sigma-Aldrich (Oakville, ON, Canada). CellTiter 96® AQueous MTS reagent powder was purchased from Promega (Madison, WI, USA). 7-AAD viability staining solution (eBioscience™) was purchased from Thermo-Fisher Scientific (Waltham, MA, USA). The remaining chemicals were obtained from Fisher Scientific (Ottawa, ON, Canada).

2.2. Microwave-Assisted Extraction

All MAE experiments were conducted using a MARS model 6® (CEM Corporation, Matthews, NC, USA) microwave-accelerated reaction system. The measurement of microwave absorptivity by various pure components was carried out using a pressure-sealed vessel by heating a given mass of compound at a specific power (800 W) and monitoring the bulk temperature. The optimization of extraction was conducted using a rotor with 75 mL Teflon pressure-sealed vessels. Grape seed powder (0.5 g) was introduced into each tube, together with 5 mL of a solution containing aqueous ethanol (26–94%). Since the maximum temperature that could be used with the present microwave tubes without ethanol leakage through the pressure release vent was 170 °C, a temperature range of 110–170 °C was used in the experiment. Accordingly, the tubes were screw-capped, shaken evenly, and placed in the rotor to allow temperature measurement by the combination of fiber optic and infrared sensor, and heated at a given temperature (110–170 °C) for 5–55 min. No stirring was applied during the heating. The temperature program consisted of a fast heating step using a fixed maximum power (800 W), followed by a plateau step during which power varied to maintain the temperature at the target value. At the end of the reaction, samples were allowed to cool down to room temperature (30 min). The solid residue was isolated by filtration. Each experiment was conducted in duplicate. The filtrate was stored at 4 °C until further analysis.

2.3. Experimental Design

Response surface methodology (RSM) was used to determine the optimal processing conditions for the MAE extraction of PAC from grape seeds powder. A central composite design (CCD) using 20 runs at low-axial, low, center, high, and high-axial levels of the three factors, namely temperature (110 °C, 120 °C, 140 °C, 160 °C, and 170 °C), ethanol concentration (26%, 40%, 60%, 80%, and 94%), and time (5 min, 15 min, 30 min, 45 min, and 55 min) was generated and analyzed using Minitab 18 software to determine the optimum settings of the factors that maximize two response variables (total monomeric catechins and total PAC content).

2.4. Extractions Using the Conventional Method

For comparison, PACs were also extracted from grape seed powder using the conventional method (extraction conditions of 94% (*v/v*) ethanol, 55 min, and 25 °C) without applying microwaves. Grape seed powder (1 g) was introduced into the tube together with 10 mL of 94% ethanol. The tube was capped and shaken at 70× *g* at room temperature for 55 min. The experiment was conducted in triplicate.

2.5. Quantification of PAC Content of Extracts

The total PAC content of the extracts was quantified using the methylcellulose precipitable (MCP) tannin assay described by Damberg et al. [19]. Methylcellulose solution (0.04% *w/v*; 1500 centipoises viscosity at 2%) was prepared. The samples and catechin standards were prepared with a 50% aqueous ethanol solution. In a 1.5 mL centrifuge tube, 300 µL of methylcellulose solution was mixed thoroughly with a 100 µL sample and 5 min was allowed for completion of the polymerization reaction. Following the addition of 200 µL of saturated ammonium sulfate solution, the sample was made up to 1 mL final volume with deionized water and the solution was vortexed. Centrifugation was performed at 10,000× *g* for 5 min. The control samples were prepared with the same volume as per the treatment sample without methylcellulose solution. After transferring 200 µL of solution from each tube into a UV-transparent 96 well plate, the absorbance at 280 nm was measured. Results were expressed in mg catechin equivalence per gram dry weight of the sample (mg CE/g DW).

2.6. UPLC-ESI-MS Analysis of Catechins in Extracts

Each extract was filtered through a 0.22 µm nylon filter and placed into amber vials. The analyses were performed using a Waters H-class UPLC separations module (Waters, Milford, MA, USA), coupled with a Quattro Micro API MS/MS system and controlled with a Masslynx V4.1 data analysis system (Micromass, Cary, NC, USA). The column used was an Aquity BEH C18 (2.1 × 100 mm, 1.7 µm) (Waters, Milford, MA, USA). For the separation of the catechin, epicatechin, epicatechin gallate (EGC), and epigallocatechin gallate (EGCG), a gradient elution was carried out with 0.1% formic acid in water (solvent A) and 0.1% formic acid in acetonitrile (solvent B) at a flow rate of 0.3 mL/min. A linear gradient profile was used, with the following proportions of solvent A applied at time t (min) (t, A%): (0, 94%), (2, 83.5%), (2.61, 83%), (2.17, 82.5%), (3.63, 82.5%), (4.08, 81.5%), (4.76, 80%), (6.75, 20%), (8.75,

94%), (12, 94%). Electrospray ionization in negative ion mode (ESI-) was used with the following conditions: capillary voltage 3000 V, cone voltage 40 V, and nebulizer gas (N_2) temperature 375 °C at a flow rate of 0.3 mL/min. Single-ion monitoring (SIM) mode using specific parent ions was employed for quantification in comparison with standards: m/z 289 for catechin and epicatechin, m/z 442 for ECG, m/z 472 for EGCG. The quantification of each analysis was performed using calibration curves created using the external standards. The limit of detection of the analytes was between 0.01 and 0.1 mg/L.

2.7. Total Antioxidant Capacity of the Extracts by FRAP Assay

The Ferric Reducing Antioxidant Power (FRAP) assay was performed as previously described by Benzie and Strain [20]. Briefly, the reagents included 300 mM acetate buffer pH 3.6, 40 mM hydrochloric acid, 10 mM 2,4,6-tripyridyl-s-triazine (TPTZ) solution, and 20 mM ferric chloride solution. The working FRAP reagent was prepared fresh on the day of analysis by mixing acetate buffer, TPTZ solution, and ferric chloride solutions in the ratio 10:1:1 and incubating at 37 °C. Absorbance was recorded at 593 nm. The calculated difference in absorbance is proportional to the ferric-reducing antioxidant power of the antioxidants present in the extracts. For quantification, a calibration curve of Trolox (0.1 mM to 1 mM) was used. The final results were expressed as mmol Trolox equivalents per L of sample. The analysis was performed in triplicate.

2.8. Inhibitory Effect of Extracts on α-Glucosidase In Vitro

The α-glucosidase-inhibitory assay was performed via the chromogenic method described by Watanabe et al. [21]. Briefly, α-glucosidase (1 U/mL, Sigma) was dissolved in 100 mM phosphate buffer (pH 6.8) containing 0.2% bovine serum albumin used as an enzyme solution. Next, 5 mM 4-nitrophenyl-a-D-glucopyranoside (PNPG) in the same buffer (pH 6.8) was used as a substrate solution. Measures of 20 μL of enzyme solution and 120 μL of strand drug/extract were mixed in a microtiter plate and incubated at 37 °C. After incubation for 15 min, the substrate solution (20 μL) was added and incubated at 37 °C for 15 min. Eighty microliters of 0.2 M sodium carbonate solution was added to stop the reaction. Absorbance was recorded at 405 nm to quantify the amount of PNP released.

2.9. Cell Culture

WRL-68 and HepG2 cells were cultured in MEME supplemented with 10% FBS, 4 mM L-glutamine, 100 U/mL penicillin, and 100 μg/mL streptomycin, as described in Thilakarathna and Rupasinghe [22]. Cell cultures were maintained at 37 °C and 5% CO_2 in a humidified incubator and subcultured before reaching confluence.

2.10. Cytotoxicity of PAC in WRL-68 and HepG2 Cells

Cytotoxicity of the PACs extracted by microwave method (MW-PAC) and conventional extraction method (Con-PACs) was evaluated in WRL-68 normal cells and HepG2 hepatocarcinoma cells to compare the cytotoxicity of PAC extracts. Cell viability (%) of the two cell lines was tested over a broad concentration range of MW-PACs and Con-PACs using the MTS cell viability assay and 7-AAD-stained flow cytometry analysis [22]. EGCG was used as a reference phenolic compound in both cell viability experiments.

2.10.1. MTS Cell Viability/Metabolic Activity Assay

Cells were seeded in a 96 well plate at a density of 6000 cells/well and incubated overnight at 37 °C and 5% CO_2 in a humidified incubator. Cells were treated with 10–1000 μg/mL concentrations of MW-PAC, Con-PAC, and EGCG for 24 h. Treated cells were exposed to MTS/PMS solution (MTS, 333 μg/mL; PMS, 25 μM of final concentration) and incubated for 3 h at 37 °C. After incubation, absorbance was measured at 490 nm using the Infinite® M200 PRO multimode microplate reader (Tecan Trading AG, Mannedorf, Switzerland).

2.10.2. 7-AAD-Stained Flow Cytometry for Cell Viability/Cell Death Evaluation

WRL-68 and HepG2 cells were seeded in a six-well plate at a density of 2×10^5 cells/well and incubated at 37 °C and 5% CO_2 in a humidified incubator overnight. Cells were treated with 10 to 1000 µg/mL concentrations of MW-PAC, Con-PAC, or EGCG for 24 h. The culture media was collected into separate tubes from each well to include potential dead cells in the flow cytometric analysis. Each well was then washed with 1 mL of PBS, and cells were harvested with TrypLE express (1 mL/well, incubated for 5 min at 37 °C). PBS (from well washing), TrypLE express, and harvested cells were pooled together with the culture media collected earlier. Samples were centrifuged at 500× g for 5 min, and the supernatant was discarded. The cell pellet was washed with cold PBS (500× g for 5 min) and re-suspended in 1 mL of PBS. Cells were stained with 5 µL of 7-AAD stain in the dark for 5 min. Samples were analyzed by flow cytometer under the FL-3 filter (BD Accuri™ C6 Plus Flow Cytometer, BD Biosciences, San Jose, CA, USA) and data were processed by Kaluza Analysis (version 2.1) FACS analysis software (Beckman Coulter Life Sciences, Indianapolis, IN, USA).

2.11. Statistical Analysis

Complete analyses of total monomeric catechins and total PAC content measured from the 20 runs (including six center points to allow the estimation of error variance) of the central composite design (CCD) conducted in random order were done. The analyses included verification that the model did not have a significant lack of fit ($p > 0.05$), which indicated the adequacy of the model to accurately predict the variation, and the normal distribution and constant variance assumptions on the error terms were valid. Independence assumption was validated through the randomization of the run order. This was followed by testing the significance of each term and constructing contour plots for each response variable to determine the best setting of the factors, and finally determining the "sweet spot" that optimized the two response variables using overlaid contour plots and the response optimizer. These statistical analyses were completed using methods described previously [23,24]. Linear regression analyses were conducted to determine the total PAC content and antioxidant capacity. Prism 6 software (GraphPad Software, La Jolla, CA, USA) was used to calculate IC_{50} values for the inhibitory effect on α-glucosidase and cell viability.

3. Results

3.1. Response Surface Modeling

3.1.1. Model Fitting of Parameters Based on Total Monomeric Catechins and Total PAC Content

A total of 20 runs were performed to optimize the three factors (temperature, ethanol concentration, and microwave extraction time) in the current CCD. Preliminary single-factor tests were conducted to determine the effective ranges for each of the three factors to be optimized. The maximum temperature that could be used with the present microwave tubes without any ethanol leakage through the pressure release vent was 170 °C. Thus, a temperature range of 110–170 °C was selected for the optimization. Reaction time was distributed around 30 min to ensure complete reactions. Ethanol concentrations ranged between 20% and 100% in order to select the optimum concentration.

The actual values of the three factors (temperature, ethanol concentration, and microwave extraction time) and experimental results for the yield of total monomeric catechins (catechin, epicatechin, and EGC) and total PACs are shown in Table 1. The obtained data were used for the prediction of an optimum set of extraction settings from grape seed powder extract with high monomeric catechins and PAC. These experimental results were fitted to a second-order response surface model, and the analysis of variance (ANOVA) p-values that show the significance of the components of the model are presented in Table 2. According to the ANOVA results, the model described the variability in total monomeric catechins and total PACs very well, with adjusted R^2 values of 85.2% and 94.2%, respectively; and without significant lack of fit ($p > 0.05$).

Table 1. Central composite design (CCD) arrangement for microwave-assisted extraction of PACs from grape seed.

Run Order	Temperature (°C)	Time (min)	Ethanol Concentration (%)	Total Monomeric Catechins (mg/g DW)	Total PACs (mg CE/g DW)
1	110 (−1.68)	30 (0)	60 (0)	11.18	108.22
2	120 (−1)	15 (−1)	40 (−1)	8.11	72.12
3	120 (−1)	15 (−1)	80 (+1)	11.72	99.92
4	120 (−1)	45 (+1)	40 (−1)	6.75	83.98
5	120 (−1)	45 (+1)	80 (+1)	11.65	104.18
6	140 (0)	30 (0)	60 (0)	11.03	106.87
7	140 (0)	30 (0)	60 (0)	9.31	107.03
8	140 (0)	30 (0)	60 (0)	9.71	106.83
9	140 (0)	30 (0)	60 (0)	9.26	107.81
10	140 (0)	30 (0)	60 (0)	8.83	103.46
11	140 (0)	30 (0)	60 (0)	9.05	107.58
12	140 (0)	30 (0)	26.4 (−1.68)	3.07	24.44
13	140 (0)	30 (0)	93.6 (+1.68)	10.64	85.48
14	140 (0)	5 (−1.68)	60 (0)	10.58	97.90
15	140 (0)	55 (+1.68)	60 (0)	9.85	90.28
16	160 (+1)	15 (−1)	40 (−1)	8.52	66.58
17	160 (+1)	15 (−1)	80 (+1)	13.84	104.64
18	160 (+1)	45 (+1)	40 (−1)	7.00	38.27
19	160 (+1)	45 (+1)	80 (+1)	15.05	90.81
20	170 (+1.68)	30 (0)	60 (0)	10.41	66.39

The CCD-coded values of temperature, time, and ethanol concentration are shown in brackets; run orders 6 to 11 represent the experiment at the center points repeated six times to allow the estimation of the error variance needed for ANOVA; total monomeric catechins included catechin, epicatechin, and EGC; PACs: proanthocyanidins; CE: catechin equivalents.

Table 2. Analysis of variance (ANOVA) p-values for the second-order response surface model for total monomeric catechins and total PAC content.

Source of Variation	Total Monomeric Catechins	Total PACs
Temp	0.179	0.001
Time	0.439	0.099
EthConc	0.001	0.001
Temp × Temp	0.020	0.005
Time × Time	0.126	0.075
EthConc × EthConc	0.015	0.001
Temp × Time	0.698	0.005
Temp × EthConc	0.112	0.026
Time × EthConc	0.181	0.683
Adjusted R^2	85.2%	94.2%

Temp × Temp, Time × Time, and EthConc × EthConc represent the quadratic components of the model; total monomeric catechins included catechin, epicatechin, and EGC; PACs: proanthocyanidins.

3.1.2. Optimization of the MAE Operating Conditions

Contour plots for yield of total monomeric catechins and total PACs are shown in Figures 1 and 2, respectively. These plots represent how total monomeric catechins and total PACs changed as microwave time and temperature changed while keeping the ethanol concentration constant at 60% (Figures 1a and 2a), as ethanol concentration and temperature changed while the microwave time was kept constant at 30 min (Figures 1b and 2b), and as ethanol concentration and time changed while keeping the temperature constant at 140 °C (Figures 1c and 2c). The strongest effect was attributed to the ethanol concentration, in agreement with the poor solubility of PACs in water. As a consequence, reducing the proportion of ethanol below 60% (v/v) decreased the amount of PACs extracted. Moreover, higher ethanol concentrations showed a significant influence on the yield of monomeric catechins from grape seed powder. Microwave temperature and duration showed some influence on the yield of monomeric catechins and PACs. Higher yields of monomeric catechins and PACs were observed when higher temperatures and microwave time were used.

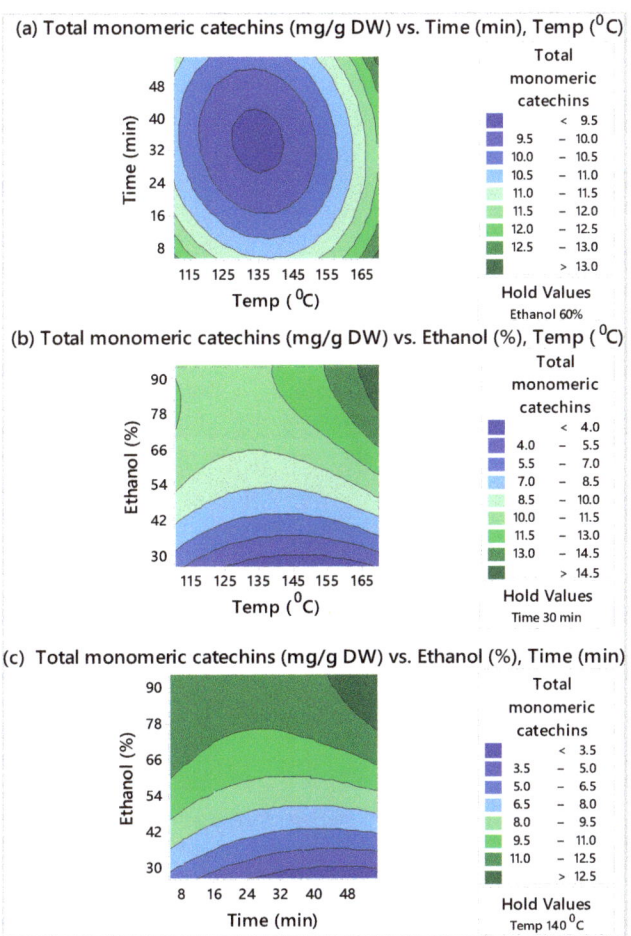

Figure 1. Contour plots of total monomeric catechins (mg/g DW) vs. (**a**) time (min) and temperature (°C) with ethanol held at 60%, (**b**) ethanol (%) and temperature (°C) with time held at 30 min, and (**c**) ethanol (%) and time (min) with temperature held at 140 °C.

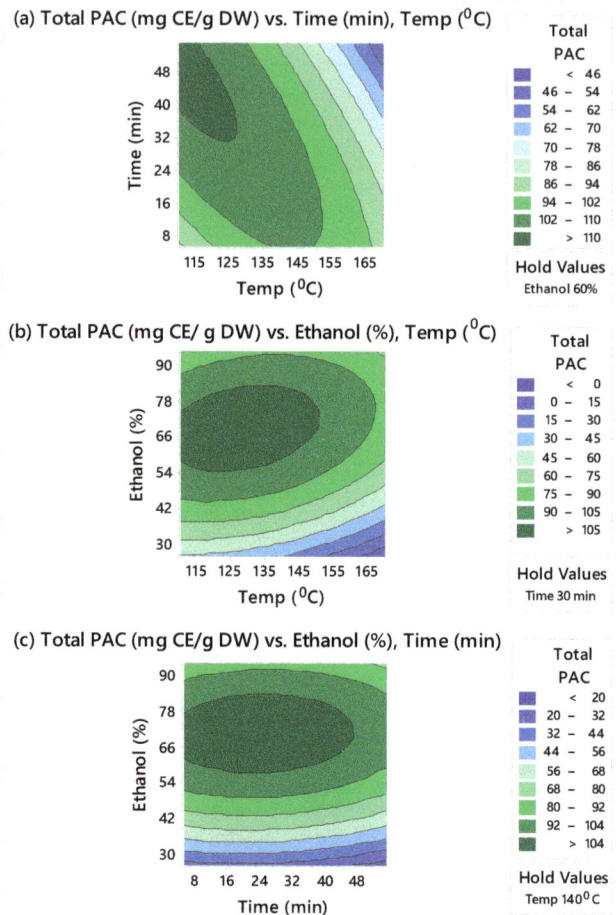

Figure 2. Contour plots of total PACs (mg CE/g DW) vs. (**a**) time (min) and temperature (C) with ethanol held at 60%, (**b**) ethanol (%) and temperature (°C) with time held at 30 min, and (**c**) ethanol (%) and time (min) with temperature held at 140 °C.

The contour plots shown in Figures 1 and 2 reflect our focus on determining the optimum settings of the factors to maximize total PACs and total monomeric catechins individually. From these figures, high values of total PACs ranged from 100 to 110 mg CE/g DW, and high values of total monomeric catechins ranged from 12 to 14 mg/g DW. Therefore, an overlaid contour plot that showed the optimum extraction conditions to maximize total monomeric catechins and total PACs jointly within these ranges was produced and is shown in Figure 3. As shown in Table 3, since the optimum times that maximized total PACs and total monomeric catechins were different (41 min and 55 min, respectively), for the overlaid plot (Figure 3), time was held at 50 min (half-way between 41 min and 55 min). Depending on what the objectives of the process are, various conditions may be preferable.

Response optimizer analysis results that show the optimum settings of temperature, ethanol concentration, and time to maximize total monomeric catechins and total PACs are shown in Table 3. Accordingly, the optimum settings for maximizing total monomeric catechins to 18.3 mg/g DW were 170 °C temperature, 94% ethanol concentration, and 55 min. However, in this setting, the predicted total PACs was 43.7 mg CE/g DW. The optimum setting for maximizing total PAC to 113.6 mg CE/g

DW was 120 °C temperature, 68% ethanol concentration, and 41 min. However, in this setting, the predicted total monomeric catechins reduced to 10.7 mg/g DW.

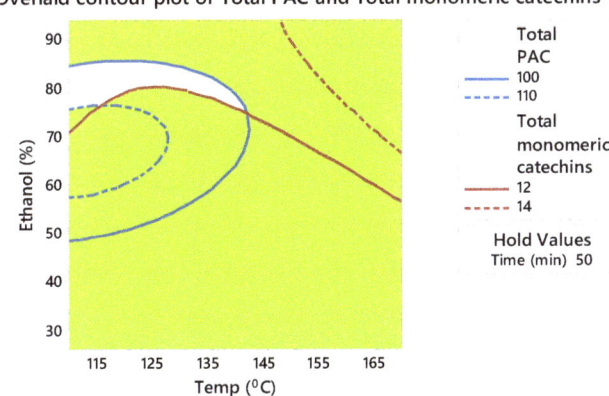

Figure 3. Overlaid contour plot showing the sweet spot (white area) for total PACs (mg CE/g DW) between 100 and 110 mg CE/g DW and total monomeric catechins (mg/g DW) between 12 and 14 mg/g DW.

Table 3. Optimum settings of temperature, ethanol concentration, and time that maximized both total monomeric catechins (mg/g DW) and total PACs (mg CE/g DW). The optimum (maximum) values were obtained by maximizing each response variable separately.

Factor	Optimum Conditions for Total Monomeric Catechins	Optimum Conditions for Total PACs
Temperature	170 °C	120.3 °C
Ethanol concentration	94%	67.9%
Time	55 min	41.4 min
Predicted concentration under the optimum condition for each analyte	18.3 mg/g DW	113.6 mg CE/g DW
Predicted concentration of total monomeric catechins under the optimum condition of total PACs	10.7 mg/g DW	n/a
Predicted concentration of total PACs under the optimum condition of total monomeric catechins	n/a	43.7 mg CE/g DW

Total monomeric catechins include catechin, epicatechin, and EGC; PACs: proanthocyanidins; CE: catechin equivalents; n/a: not applicable; DW, dry weight.

3.2. Comparison with Conventional Extraction

3.2.1. Quantitative Measurement

To determine whether microwaves played an important role in the reactivity and selectivity of the process, we compared the yields and characteristics of the extracts obtained using microwave irradiation (MW) and conventional extraction (Con) under similar conditions: 94% EtOH; shaking at 100× g, 55 min; 170 °C for MW and 25 °C for Con. The yields of monomeric catechins and total PAC from grape seed by MW (MW-PAC) and from grape seed powder by Con (Con-PAC) are presented in Figure 4A,B. Total monomeric catechins of Con-PAC consisted of catechin, epicatechin, and EGC at 0.95, 1.04, and 0.11 mg/g DW, respectively. Under conventional conditions (94% EtOH; 25 °C; 55 min), the values for total monomeric catechins and PAC were 2.10 ± 0.13 mg/g DW and 9.70 ± 0.39 mg CE/g DW, respectively. Total monomeric catechins of MW-PAC consisted of catechin, epicatechin, and EGC at 4.0, 3.32, and 0.83 mg/g DW, respectively. Under microwave extraction conditions (94% ethanol; 170 °C; 55 min), the values for total monomeric catechins and PAC were 8.15 ± 0.20 mg/g DW and 56.37 ± 8.37 mg CE/g DW, respectively.

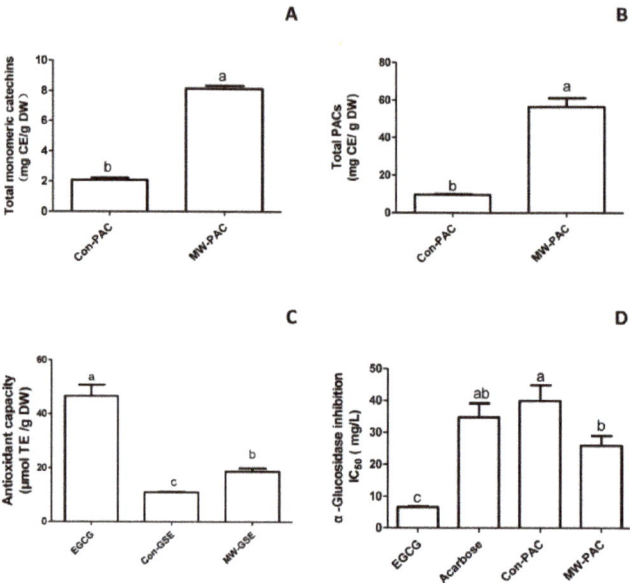

Figure 4. The total monomeric catechins (**A**), total PACs (**B**), total antioxidant capacity (FRAP method) (**C**), and inhibition of α-glucosidase (**D**) of Con-PAC and MW-PAC. The total PAC content is expressed in mg catechin equivalence (CE)/g DW of the sample. Epigallocatechin gallate (EGCG), and acarbose were used for the comparison purpose. Means sharing the same letter are not significantly different at the 5% level.

3.2.2. Total Antioxidant Capacity

Accumulating evidence demonstrates that flavanols, including catechins and PACs, which have high antioxidant capacity, help to decrease the risk of developing chronic diseases including cancer, cardiovascular disease, and diabetes [12,25,26]. Therefore, plant-based therapeutics present potential alternative therapies that should be explored due to their reported safety and health benefits. A FRAP assay was performed to compare the total antioxidant capacity of the extracts by microwave irradiation (MW-PAC) and conventional extraction (Con-PAC). EGCG was used as a reference compound (Figure 4C). MW-PAC (186.9 ± 12.3 μmol TE/L) had greater FRAP values than Con-PAC (109.6 ± 2.4 μmol TE/L). Results indicated that the overall antioxidant capacity was higher in MW-PAC as compared to the corresponding Con-PAC. Pearson's correlation coefficient was calculated to evaluate the relationship between total PACs and antioxidant capacity. The total PACs of Con-PAC and MW-PAC all showed significant linear correlation (r = 0.9984; $p < 0.0001$, r = 0.9984; $p < 0.0001$, respectively) with FRAP values.

3.2.3. α-Glucosidase-Inhibitory Activity

EGCG, acarbose (the drug), Con-PAC, and MW-PAC were tested for their ability to inhibit α-glucosidase activity (Figure 4D). Acarbose showed an IC_{50} value of 347.4 ± 43.5 mg/L. EGCG exhibited the highest inhibitory activity with the lowest IC_{50} value of 66.9 ± 1.6 mg/L. Compare to the acarbose, Con-PAC and MW-PAC showed comparable inhibitory activities against α-glucosidase (IC_{50}: 399.1 ± 48.9 mg/L and 259.0 ± 30.8 mg/L, respectively).

3.2.4. Cytotoxicity of Con-PAC and MW-PAC in WRL-68 and HepG2 Cells

The potential anticancer activity of MW-PAC and Con-PAC was measured by the means of capacity to kill hepatocarcinoma (HepG2) cells in comparison to normal (non-malignant) hepatic (WRL-68) cells. The Con-PAC-, MW-PAC-, and EGCG-mediated cytotoxicity were concentration-dependent for both cell lines (Figures 5 and 6). Con-PAC exhibited significant toxicity in both cell lines at concentrations of 250 µg/mL and higher. Cytotoxicity of MW-PAC in both cell lines was similar to that of the Con-PAC, causing higher toxicity at the 250 µg/mL concentration level. The EGCG-mediated cytotoxicity was intense compared to the Con-PAC and MW-PAC. A drastic reduction in % cell viability was observed for EGCG in both cell lines at the concentration of 100 µg/mL. A beneficial effect of selective cytotoxicity in hepatocarcinoma cells (HepG2) compared to healthy hepatic cells (WRL 68) was not observed for either Con-PAC or MW-PAC.

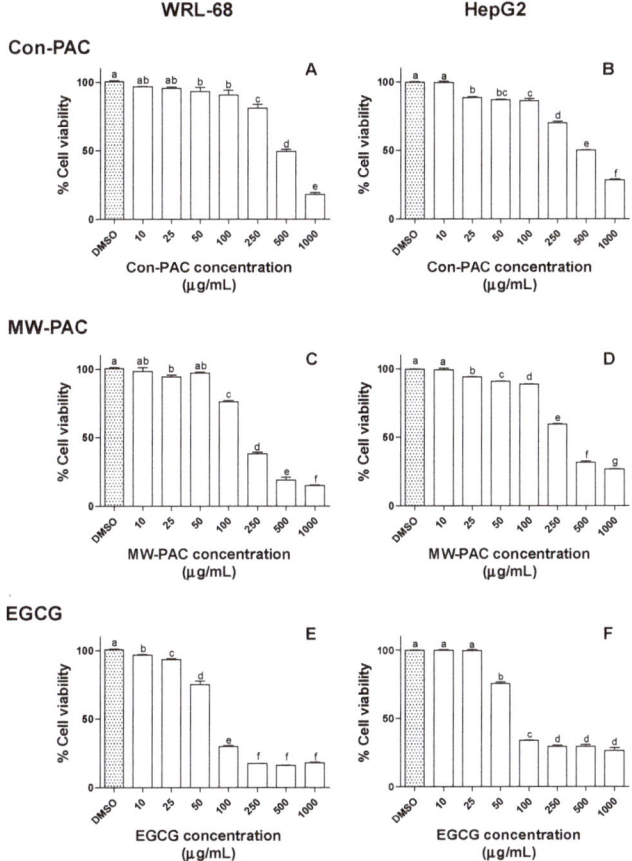

Figure 5. Cell viability (%) of WRL-68 and HepG2 cells exposed to different concentrations of Con-PAC, MW-PAC, and EGCG measured by MTS cell viability assay. WRL-68 and HepG2 cells were exposed to 10–1000 µg/mL concentrations of Con-PAC (**A** and **B**, respectively), MW-PAC (**C** and **D**, respectively), and EGCG (**E** and **F**, respectively) for 24 h. Cell viability of the WRL-68 and HepG2 cells was measured by incubating (3 h at 37 °C) the cells with MTS cell viability reagent. Results presented are means ± SD of three independent experiments. Means sharing the same letter are not significantly different at the 5% level. Con-PAC: PACs extracted by conventional method; MW-PAC: PACs extracted by microwave-assisted method; EGCG, epigallocatechin gallate.

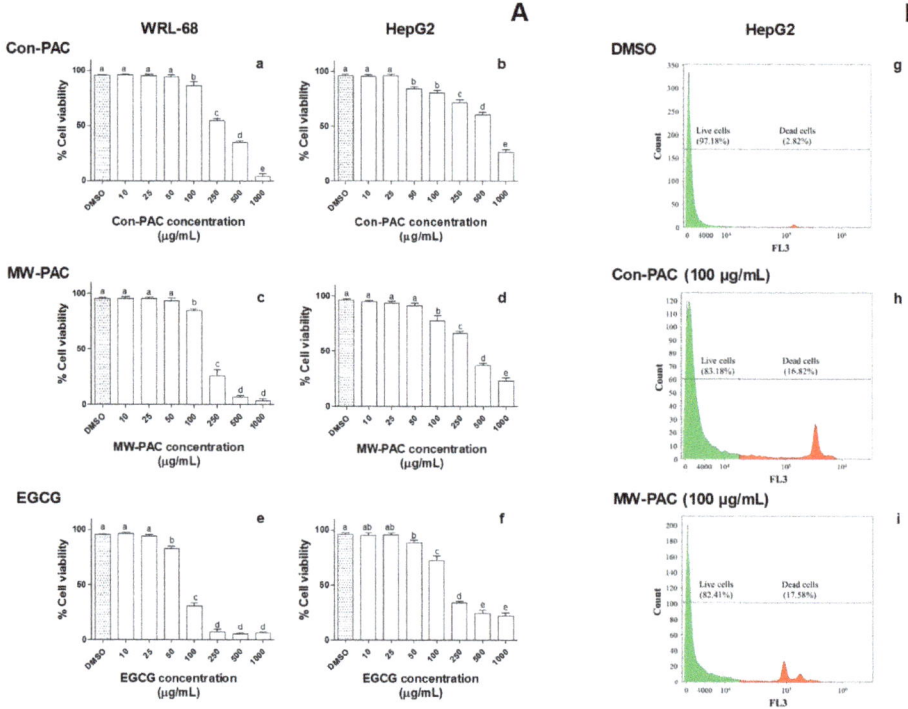

Figure 6. Cell viability (%) of WRL-68 and HepG2 cells exposed to different concentrations of Con-PAC, MW-PAC, and EGCG measured by 7-AAD-stained flow cytometry (**A**) and FL3 histograms of the HepG2 cells (**B**) treated with 100 μg/mL of PAC. WRL-68 and HepG2 cells were exposed to 10–1000 μg/mL concentrations of Con-PAC, MW-PAC, and EGCG for 24 h. Cells were stained with 7-AAD stain in the dark (5 min) and analyzed by flow cytometry under the FL-3 filter. FACS data were processed by Kaluza Analysis (version 2.1) flow cytometry analysis software. Results presented are means ± SD of three independent experiments. Means sharing the same letter are not significantly different at the 5% level. PACs: proanthocyanidins; Con-PAC: proanthocyanidins extracted by the conventional method; MW-PAC: proanthocyanidins extracted by microwave-assisted method; EGCG, epigallocatechin gallate.

4. Discussion

Wine production residue has been the subject of extensive research due to its high content of bioactives with important biological functions. Among the different wine industry byproducts, grape seeds contain the highest amount of total phenolic compounds. Catechins and their isomers and polymers are the main phenolic components in grape seeds [27]. Due to their natural antioxidant abilities, GSE can be employed as a functional ingredient in value-added food and nutraceutical products. However, the biological activities of PACs from grape seeds are often confounded by their seemingly low bioavailability. In the present study, MAE was successfully used to extract PACs from grape seeds with enhanced monomeric catechins (increased from 2.10 ± 0.13 mg/g DW to 8.15 ± 0.20 mg/g DW). Advanced graphical and numerical optimizations were run to determine the optimum levels of studied extraction conditions with desirable levels of monomeric catechins and PAC. Under the optimum settings (170 °C; 94% ethanol; 55 min) for maximizing total monomeric catechins, replacing conventional extraction by MAE led to a higher yield of PACs (5.8-fold greater than Con) and monomeric catechins (3.9-fold greater than Con). The higher yields of monomeric catechins obtained for MW-PAC were likely due to extensive PAC depolymerization under the MAE conditions. Higher yields of PACs have also been observed by others [17,18] when combining an MW pretreatment with two-phase aqueous extraction of polyphenols from grape seeds. However, the experimental value for total monomeric catechins (8.15 ± 0.20 mg/g) was lower than the predicted value (18.3 mg/g DW), and the experimental value for total PAC (56.37 ± 8.37 mg CE/g DW) was higher than the predicted value (43.7 mg CE/g DW). Recently, the monograph of green tea was included in the European Pharmacopoeia, wherein green tea is standardized for caffeine content (min 1.5%) and for the total content of catechins, expressed as EGCG (min 8%) [28]. The total monomeric catechin content of the extract by MW in this study was over this standard. As far as we are aware, this is the first study to report microwave irradiation to increase the yields of monomeric catechins through depolymerization of the PACs of grape seeds.

Previous studies have suggested that the content of flavanols from the diet could significantly contribute to their total antioxidant capacities [29]. In the current study, the antioxidant potential of the GSE obtained by MAE and conventional extraction methods were compared. MW-PAC showed higher antioxidant capacity than that of Con-PAC. The correlation analysis showed a significant linear correlation between PAC content and FRAP, indicating that the flavanols in GSE were substantially responsible for the antioxidant capacity.

Dietary flavanols, in addition to their antioxidant effects, have been reported to exert anti-hyperglycemic effects by binding to glucose transporters and competitive inhibition of digestive enzymes [30]. α-Glucosidase is one of the important carbohydrate-hydrolyzing enzymes that digest dietary starch and degrade oligosaccharides to glucose, resulting in postprandial glucose surge. Therefore, inhibition of α-glucosidase activities is one of the primary approaches used to manage hyperglycemic conditions of type 2 diabetic patients. The possibility of clinical use of such inhibitors for diabetic or obese patients has been suggested using acarbose, which has been shown to effectively reduce the intestinal absorption of sugars in humans. MW-PAC exhibited greater α-glucosidase inhibition (IC_{50} of 259.0 ± 30.8 mg/L) than Con-PAC (IC_{50} of 399.1 ± 48.9 mg/L), suggesting that depolymerization of PACs to monomeric catechins enhanced their biological activity. This observation also agrees with the reports that polyphenols are effective inhibitors of α-glucosidase [31]. Interestingly, the activities of both MW-PAC and Con-PAC were comparable to the drug acarbose's IC_{50} value (347.4 ± 43.5 mg/L).

The cytotoxicity of Con-PAC and MW-PAC in HepG2 and WRL-68 cells was evaluated. Both Con-PAC and MW-PAC showed similar cytotoxicity in HepG2 and WRL-68 cells. Con-PAC and MW-PAC did not cause relatively higher toxicity in HepG2 cancer cells over WRL-68 normal cells. However, further studies can be recommended to compare the biological activities of MW-PAC at different degrees of depolymerization to prevent or reduce cancer initiation and progression. The high cytotoxicity of EGCG compared to the extracted PACs could have been due to the formation of H_2O_2 in the cell culture media [32].

5. Conclusions

Overall, this is the first report of an application of microwaves as the energy source for achieving PAC depolymerization. The present results showed that MAE is more effective than the conventional method for recovering monomeric catechins and PACs from grape seed powder. PACs are partially depolymerized under the conditions of high ethanol concentration (94%), high temperature (170 °C), and 55 min extraction. The product of MAE showed higher antioxidant capacity and α-glucosidase-inhibitory activities than that of the conventional method, due to the increase of monomeric catechins. The present findings suggest the potential use of industrial byproducts from grape-processing industries as a renewable resource of functional food ingredients and nutraceuticals.

Author Contributions: Conceptualization, H.P.V.R.; methodology, J.C., W.P.D.W.T., T.A.; software and formal analysis, T.A.; investigation, J.C., W.P.D.W.T., T.A., H.P.V.R.; resources and data curation, H.P.V.R.; writing—original draft preparation, J.C., W.P.D.W.T., T.A.; writing—review and editing, H.P.V.R.; visualization, J.C., W.P.D.W.T., T.A.; supervision, H.P.V.R., T.A.; project administration, H.P.V.R.; funding acquisition, H.P.V.R., J.C. All authors have read and agreed to the published version of the manuscript.

Funding: This research was funded by the China Scholarship Council (JC, grant number, 201606275184) and Natural Sciences and Engineering Research Council (NSERC) of Canada (HPVR, grant number RGPIN/04594-2017) and the APC was funded by HPVR and JC.

Acknowledgments: The authors wish to thank Royal Grapeseed, Milton, NY, USA for providing the grape seed used for this research.

Conflicts of Interest: The authors declare no conflict of interest. The funders had no role in the design of the study; in the collection, analyses, or interpretation of data; in the writing of the manuscript, or in the decision to publish the results.

References

1. Borges, G.; Ottaviani, J.I.; van der Hooft, J.J.J.; Schroeter, H.; Crozier, A. Absorption, metabolism, distribution and excretion of (−)-epicatechin: A review of recent findings. *Mol. Asp. Med.* **2018**, *61*, 18–30. [CrossRef]
2. Fan, F.Y.; Sang, L.X.; Jiang, M. Catechins and their therapeutic benefits to inflammatory bowel disease. *Molecules* **2017**, *22*, 484. [CrossRef]
3. Fathima, A.; Rao, J.R. Selective toxicity of catechin—A natural flavonoid towards bacteria. *Appl. Microbiol. Biotechnol.* **2016**, *100*, 6395–6402. [CrossRef]
4. Anantharaju, P.G.; Gowda, P.C.; Vimalambike, M.G.; Madhunapantula, S.V. An overview on the role of dietary phenolics for the treatment of cancers. *Nutr. J.* **2016**, *15*, 99. [CrossRef]
5. Chang, H.P.; Sheen, L.Y.; Lei, Y.P. The protective role of carotenoids and polyphenols in patients with head and neck cancer. *J. Chin. Med. Assoc.* **2015**, *78*, 89–95. [CrossRef]
6. Chen, W.; Becker, T.; Qian, F.; Ring, J. Beer and beer compounds: Physiological effects on skin health. *J. Eur. Acad. Dermatol. Venereol.* **2014**, *28*, 142–150. [CrossRef]
7. Losada-Echeberria, M.; Herranz-Lopez, M.; Micol, V.; Barrajon-Catalan, E. Polyphenols as promising drugs against main breast cancer signatures. *Antioxidants* **2017**, *6*, 88. [CrossRef] [PubMed]
8. Moga, M.A.; Dimienescu, O.G.; Arvatescu, C.A.; Mironescu, A.; Dracea, L.; Ples, L. The role of natural polyphenols in the prevention and treatment of cervical cancer—An overview. *Molecules* **2016**, *21*, 1055. [CrossRef] [PubMed]
9. Vezza, T.; Rodriguez-Nogales, A.; Algieri, F.; Utrilla, M.P.; Rodriguez-Cabezas, M.E.; Galvez, J. Flavonoids in inflammatory bowel disease: A review. *Nutrients* **2016**, *8*, 211. [CrossRef]
10. Martin, M.A.; Goya, L.; Ramos, S. Potential for preventive effects of cocoa and cocoa polyphenols in cancer. *Food Chem. Toxicol.* **2013**, *56*, 336–351. [CrossRef] [PubMed]
11. Zhao, Y.; Hu, X.; Zuo, X.; Wang, M. Chemopreventive effects of some popular phytochemicals on human colon cancer: A review. *Food Funct.* **2018**, *9*, 4548–4568. [CrossRef] [PubMed]
12. Parmar, I.; Rupasinghe, H.P.V. Proanthocyanidins in cranberry and grape seeds: Metabolism, bioavailability and biological activity. In *Nutraceuticals and Functional Foods: Natural Remedy*; Brar, S.K., Kaur, S., Dhillon, G.S., Eds.; Nova Science Publishers: Hauppauge, NY, USA, 2014; pp. 119–145.
13. Bernatoniene, J.; Kopustinskiene, D.M. The role of catechins in cellular responses to oxidative stress. *Molecules* **2018**, *23*, 965. [CrossRef] [PubMed]
14. Ma, Z.F.; Zhang, H. Phytochemical constituents, health benefits, and industrial applications of grape seeds: A mini-review. *Antioxidants* **2017**, *6*, 71. [CrossRef] [PubMed]
15. Bizzi, C.A.; Pedrotti, M.F.; Silva, J.S.; Barin, J.S.; Nóbrega, J.A.; Flores, E.M.M. Microwave-assisted digestion methods: Towards greener approaches for plasma-based analytical techniques. *J. Anal. At. Spectrom.* **2017**, *32*, 1448–1466. [CrossRef]
16. Radojković, M.; Moreira, M.M.; Soares, C.; Fátima Barroso, M.; Cvetanović, A.; Švarc-Gajić, J.; Morais, S.; Delerue-Matos, C. Microwave-assisted extraction of phenolic compounds from *Morus nigrab* leaves: Optimization and characterization of the antioxidant activity and phenolic composition. *J. Chem. Technol. Biotechnol.* **2018**, *93*, 1684–1693. [CrossRef]

17. Dang, Y.-Y.; Zhang, H.; Xiu, Z.-L. Microwave-assisted aqueous two-phase extraction of phenolics from grape (Vitis vinifera) seed. *J. Chem. Technol. Biotechnol.* **2014**, *89*, 1576–1581. [CrossRef]
18. Romero-Diez, R.; Matos, M.; Rodrigues, L.; Bronze, M.R.; Rodriguez-Rojo, S.; Cocero, M.J.; Matias, A.A. Microwave and ultrasound pre-treatments to enhance anthocyanins extraction from different wine lees. *Food Chem.* **2019**, *272*, 258–266. [CrossRef]
19. Dambergs, R.G.; Mercurio, M.D.; Kassara, S.; Cozzolino, D.; Smith, P.A. Rapid measurement of methyl cellulose precipitable tannins using ultraviolet spectroscopy with chemometrics: Application to red wine and inter-laboratory calibration transfer. *Appl. Spectrosc.* **2012**, *66*, 656–664. [CrossRef]
20. Benzie, I.F.S.; Strain, J.J. Ferric reducing/antioxidant power assay: Direct measure of total antioxidant activity of biological fluids and modified version for simultaneous measurement of total antioxidant power and ascorbic acid concentration. *Methods Enzymol.* **1999**, *299*, 15–27.
21. Watanabe, J.; Kawabata, J.; Kurihara, H.; Niki, R. Isolation and identification of α-glucosidase inhibitors from Tochu-cha (*Eucommia ulmoides*). *Biosci. Biotechnol. Biochem.* **2014**, *61*, 177–178. [CrossRef]
22. Thilakarathna, W.P.D.W.; Rupasinghe, H.P.V. Microbial metabolites of proanthocyanidins reduce chemical carcinogen-induced DNA damage in human lung epithelial and fetal hepatic cells in vitro. *Food Chem. Toxicol.* **2019**, *125*, 479–493. [CrossRef] [PubMed]
23. Myers, R.H.; Montgomery, D.C.; Anderson-Cook, C.M. *Response Surface Methodology: Process and Product Optimization Using Designed Experiments*, 4th ed.; Wiley: New York, NY, USA, 2016.
24. Montgomery, D.C. *Design and Analysis of Experiments*, 9th ed.; Wiley: New York, NY, USA, 2017.
25. Majewska, M.; Lewandowska, U. The chemopreventive and anticancer potential against colorectal cancer of polyphenol-rich fruit extracts. *Food Rev. Int.* **2017**, *34*, 390–409. [CrossRef]
26. Ramos, S.; Martin, M.A.; Goya, L. Effects of cocoa antioxidants in type 2 diabetes mellitus. *Antioxidants* **2017**, *6*, 84. [CrossRef] [PubMed]
27. Yammine, S.; Brianceau, S.; Manteau, S.; Turk, M.; Ghidossi, R.; Vorobiev, E.; Mietton-Peuchot, M. Extraction and purification of high added value compounds from by-products of the winemaking chain using alternative/nonconventional processes/technologies. *Crit. Rev. Food Sci. Nutr.* **2018**, *58*, 1375–1390. [CrossRef]
28. Council of Europe. Green tea (*Camelliae sinensis* non fermentata folia). In *European Pharmacopoeia*, 9th ed.; Council of Europe: Strasbourg, France, 2018.
29. Valls, J.; Agnolet, S.; Haas, F.; Struffi, I.; Ciesa, F.; Robatscher, P.; Oberhuber, M. Valorization of Lagrein grape pomace as a source of phenolic compounds: Analysis of the contents of anthocyanins, flavanols and antioxidant activity. *Eur. Food Res. Technol.* **2017**, *243*, 2211–2224. [CrossRef]
30. Sekhon-Loodu, S.; Rupasinghe, H.P.V. Evaluation of antioxidant, antidiabetic and antiobesity potential of selected traditional medicinal plants. *Front. Nutr.* **2019**, *6*, 53. [CrossRef]
31. Wang, H.; Liu, T.; Huang, D. Starch hydrolase inhibitors from edible plants. *Adv. Food Nutr. Res.* **2013**, *70*, 103–136.
32. Weisburg, J.H.; Weissman, D.B.; Sedaghat, T.; Babich, H. In vitro cytotoxicity of epigallocatechin gallate and tea extracts to cancerous and normal cells from the human oral cavity. *Basic Clin. Pharmacol. Toxicol.* **2004**, *95*, 191–200. [CrossRef]

 © 2020 by the authors. Licensee MDPI, Basel, Switzerland. This article is an open access article distributed under the terms and conditions of the Creative Commons Attribution (CC BY) license (http://creativecommons.org/licenses/by/4.0/).

Article

Preliminary Studies on the Application of Grape Seed Extract in the Dyeing and Functional Modification of Cotton Fabric

Ling Guo [1], Zhi-Yi Yang [2], Ren-Cheng Tang [1,*] and Hua-Bin Yuan [1]

1. National Engineering Laboratory for Modern Silk, College of Textile and Clothing Engineering, Soochow University, 199 Renai Road, Suzhou 215123, China; Emily26@live.cn (L.G.); huabinyuan19@163.com (H.-B.Y.)
2. Lushan College, Guangxi University of Science and Technology, Liuzhou 545000, China; zhiyiyang666@163.com
* Correspondence: tangrencheng@suda.edu.cn; Tel.: +86-512-6716-4993

Received: 31 December 2019; Accepted: 30 January 2020; Published: 2 February 2020

Abstract: Cotton has the shortcomings of having no antibacterial, antioxidant and ultraviolet (UV) protection properties, which are of great importance for health protection purposes. In the present study, grape seed extract (GSE) mainly composed of proanthocyanins (tannins) was employed to simultaneously import pale colors and the three aforementioned functions to cotton fabric. The tests on the application conditions of GSE showed that pH and GSE concentration had great impact on the color depth of cotton fabric, and the color hue of dyed fabric could be controlled in the absence of pH regulators due to the weakly acidic nature of GSE solution. The fabric dyed with 10%owf (on the weight of fabric) GSE exhibited an excellent inhibition effect towards *Escherichia coli*, whereas the one dyed with 20%owf GSE had high antioxidant activity of 97%. The fabric dyed with 5%owf GSE offered excellent UV protection. This study reveals that GSE can be used as a functional finishing agent for health protection in cotton textiles in addition to coloration capability.

Keywords: grape seed extract; tannins; cotton; antibacterial; antioxidant; UV protection

---

1. Introduction

Tannins are the second-largest aromatic compounds in nature and are present in various tissues of many higher plants [1]. Tannins are classified into two categories: hydrolyzable and condensed. The latter occupies a prominent position in tannin products sold globally, whereas the former accounts for about 10% [1,2]. Hydrolyzable tannins consist of gallic acid esters (gallotannins) linked to glucose or polymerized forms of hexahydroxydiphenol (ellagitannins). Condensed tannins are oligomers composed of 3–8 repetitive units of flavonoid [1].

The earliest applications are found in the field of leather and textile processing. In ancient times, tannins were used as a tanning material of leathers [3], and a colorant for the black coloring of silk, wool and cotton textiles, where the black colors result from the complexation between tannins and iron salts [4–7]. Later, more applications of tannins were further reported. Nowadays, tannins are widely used in leather tanning, wood adhesives, pharmaceuticals and medicines, additives of wine and fruit juices, flocculants of polluting materials, inhibitors of corrosion of metals, plastic resins, preservatives, and additives of flame retardant and insulating materials [3]. In recent years, the use of natural dyes as substitutes of some synthetic dyes is on the rise due to their environmental biocompatibility. Thus, tannin dyes are taken seriously again, considering their historical position in the dyeing of ancient textiles. In addition to normal coloring, the antibacterial, antioxidant, ultraviolet (UV) protection, and flame retardant properties of tannins on textiles are being utilized to develop functional textiles [7–11].

Grapes are one of the largest fruit crops all over the world. A high proportion (about 80%) of grapes is applied to winemaking, whereas others are consumed as raisins, table grapes, and juices [12,13]. In the process of winemaking, solid wastes accounting for about 25–30% of grapes are produced, which include grape skins and seeds [13,14]. These wastes, being the cheap source of natural phenolic compounds, contain phenolic acids (e.g., gallic acid), catechin, anthocyanins, and simple and complex flavonoids (e.g., proanthocyanidins) [15–19]. Due to their good bioactivities and functions [20], the extracts from the byproducts are widely used in the food, food packaging, biomedicine, and cosmetics industries [18,19]. Several researchers have reported the application of bio-colorants derived from grape pomace in the dyeing of textiles [21–23]. Grape pomace colorants exhibited good dyeing properties for wool fabric with accepted color-fastness, and the color depth of dyed wool depended greatly on pH and temperature [21]. Acrylic fibers could be dyed with grape pomace colorants after acrylic modification by cationization and amidoximation [22]. In another study, the extraction from fermented grape seeds, skin, and stem was employed to dye cotton fabric, and the resulting color was light reddish-brown [23]. However, the functions of these dyed textiles were not reported.

Proanthocyanidins (condensed tannins) are essential grape seed constituents [20], and consist of a series of polymerized flavan-3-ols which are linked principally through the 4 and the 8 positions, as shown in Figure 1 [24,25]. In consideration of previous studies on the application of tannins in textile processing, the present study aims at exploring the application of grape seed extract (GSE) containing 95% proanthocyanins in the dyeing and functional modification of cotton textile. For this purpose, the stability of the GSE solutions at different pH values and two temperatures, the dyeing conditions of GSE, and the color stability and fastness of GSE-dyed fabrics were firstly tested. Subsequently, the antibacterial, antioxidant and UV protective properties of the fabrics dyed with different concentrations of GSE were evaluated.

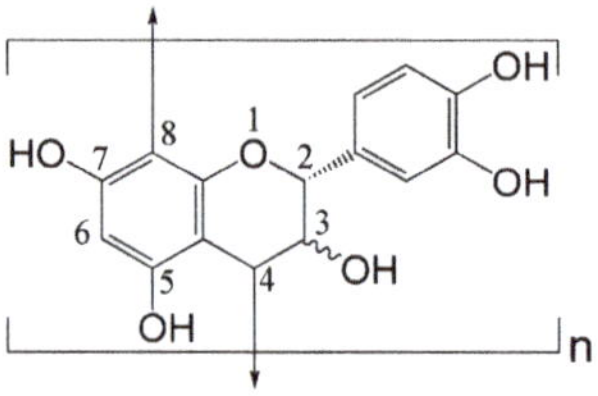

Figure 1. Representative chemical structure of flavan-3-ol unit and proanthocyanidins.

2. Materials and Methods

2.1. Materials

The scoured and bleached knitted cotton fabric was kindly provided by Longsheng Knitting Printing and Dyeing Co.Ltd., Jinjiang, Fujian Province, China. Grape seed extract (GSE) containing 95% proanthocyanins was purchased from Xi'an Huike Biological Co., Ltd., China. 2,2′-Azino-bis (3-ethylbenzothiazoline-6-sulphonic acid) diammonium salt (ABTS) was bought from Sigma-Aldrich (Shanghai) Trading Co. Ltd., China. Nutrient agar and nutrient broth were obtained from Sinopharm Chemical Reagent Co. Ltd., Shanghai, China, and Shanghai Sincere Biotech Co. Ltd., Shanghai, China, respectively. Citric acid, disodium hydrogen phosphate, sodium dihydrogen phosphate, potassium dihydrogen phosphate, sodium carbonate, and potassium persulfate were of analytical reagent grade.

2.2. Dyeing of Cotton Fabric with GSE

Cotton fabric was dipped into the GSE solution at a 50:1 bath ratio (the ratio of liquor volume to fabric weight). The dyeing was started at 20 °C, and then the solution was heated to the required temperature at a rate of 3 °C/min. At this temperature, the dyeing was continuously conducted for a set time. Afterwards, the fabric was taken out, washed thoroughly in water and air-dried.

The four dyeing conditions of GSE, including pH, temperature, time, and GSE concentration, were studied. The detailed experimental conditions are listed in Table 1. In the experiment of pH effect, the buffer consisting of citric acid and disodium hydrogen phosphate was used to adjust the pH of the GSE solution.

Table 1. Application conditions of grape seed extract (GSE).

Varible	Levels	Other Parameters
pH	4–8	GSE 20%owf, 90 °C, 60 min
Temperature	60–90 °C	GSE 20%owf, 60 min
Time	30–90 min	GSE 20%owf, 90 °C
GSE concentration	5–60% owf	90 °C, 60 min

Note: owf = on the weight of fabric.

2.3. Measurements

The ultraviolet–visible (UV–vis) absorption spectra of GSE solution (0.1 g/L) at different pH values were measured using the Shimadzu UV-1800 UV–vis spectrophotometer (Shimadzu Co., Kyoto, Japan). For this study, citric acid/disodium hydrogen phosphate buffer, sodium dihydrogen phosphate/disodium hydrogen phosphate buffer, and sodium carbonate were used to adjust the pH to 4.3, 6.7, and 10.5, respectively. GSE solution was heated to 50 °C and 90 °C at a rate of 3 °C/min, and the two temperatures were kept for 60 min. Later, the GSE solution was cooled, and the spectrophotometric analysis was carried out.

The color parameters of cotton fabric were measured by the HunterLab UltraScan PRO reflectance spectrophotometer (Hunter Associates Laboratory Inc., Reston, VA, USA) using the D65 illuminant and 10 °C standard observer. Each sample was folded two times so as to get four layers, and the average of four to six measurements was reported. The following parameters were used to evaluate the colors of dyed fabric: apparent color strength (K/S), color difference (ΔE), lightness (L*), redness–greenness index (a*), yellowness–blueness index (b*), and chroma (C*) [26]. The color difference between the undyed and dyed fabrics was calculated by Equation (1) [26]:

$$\Delta E = \sqrt{\left(L_2^* - L_1^*\right)^2 + \left(a_2^* - a_1^*\right)^2 + \left(b_2^* - b_1^*\right)^2} \tag{1}$$

where subscripts 2 and 1 denote the dyed and undyed fabrics, respectively.

The washing and rubbing color fastness of cotton fabrics dyed with 10% and 20%owf GSE were evaluated according to ISO 105-C06 and ISO 105-X12, respectively; for the two tests, a WashTec-P fastness tester (Roaches International, England) and a Model 670 crockmaster (James H. Heal, England) were used.

The antibacterial activity of cotton fabrics dyed without and with 5% and 10%owf GSE against *Escherichia coli* (*E. coli*) was evaluated with reference to the Chinese national standard GB/T 20944.3-2008 [27]. The fabric pieces were soaked into the conical flasks containing bacterial solutions, and then the solutions were shaken at 30 °C for 24 h. Afterwards, the bacterial solutions were diluted with sterilizing phosphoric buffer. The diluted bacterial solutions were inoculated onto the petri dishes, and cultured at 37 °C for 24 h. After bacterial culture, the bacterial colonies propagating on the petri dishes were photographed and recorded, and the antibacterial activity was calculated by Equation (2):

$$\text{Antibacterial activity (\%)} = 100 \cdot \frac{N_{\text{undyed}} - N_{\text{dyed}}}{N_{\text{undyed}}} \tag{2}$$

where N_{undyed} and N_{dyed} are the quantity of the visual bacterial colonies for the undyed and dyed fabrics, respectively.

The antioxidant activity of dyed cotton fabrics was evaluated by the ABTS radical decolorization method [28,29]. Firstly, the ABTS radical cation (ABTS$^+$) solution was prepared by means of the reaction between ABTS (7 mM) solution and potassium persulfate (2.45 mM) and kept in the dark for about 15 h at room temperature. Prior to use, the ABTS$^+$ solution was diluted with phosphate buffer (0.1 M, pH 7.4) to reach an absorbance of 0.700 ± 0.025 at 734 nm. Later, 10 mg of the fabric sample was soaked into 10 mL of the ABTS$^+$ solution. After 30 min, the decolorization extent of the ABTS$^+$ solution was studied by the spectrophotometric measurement. A high decolorization extent of the ABTS$^+$ solution represents a high capability of the fabric to scavenge ABTS$^+$, i.e., a high antioxidant activity. The decolorization rate of the ABTS$^+$ solution was calculated by Equation (3):

$$\text{Decolorization rate of ABTS cation } (\%) = 100 \cdot \frac{A_1 - A_2}{A_1} \tag{3}$$

where A_1 and A_2 are the absorbance of the ABTS$^+$ solution before and after decolorization, respectively.

The UV protection factor (UPF) and UV transmittance of cotton fabric were determined by the Labsphere UV-1000F ultraviolet transmittance analyzer (Labsphere Inc., North Sutton, NH, USA); a single layer of the sample was tested at five different positions, and the average of the data was reported.

3. Results and Discussion

3.1. UV–Vis Absorption Spectroscopic Study of GSE

The main component of GSE used in this study is proanthocyanins. In an aqueous solution, phenolic hydroxyl groups in proanthocyanins easily suffer from ionization and oxidation, which can change the water solubility and stability of GSE. Thus, the spectrophotometric study of the GSE solution was carried out. Figure 2 shows the UV–vis absorption spectra of the GSE solution at different pHs, which were subjected to heat treatment at 50 and 90 °C. In the case of no addition of pH regulator, the pH of the GSE solution was 5.1 due to the ionization of phenolic hydroxyl groups in proanthocyanins. Under weakly acidic conditions (pH 4.3 (buffer) and no pH regulator), the GSE solution had a strong absorption band at 280 nm and an inflection point minimum at 258 nm. Such absorption features confirm that the main component of GSE is condensed tannins [30].

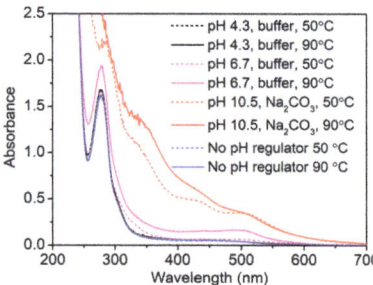

Figure 2. UV–vis absorption spectra of GSE solution at different pHs and two temperatures (GSE concentration, 0.1 g/L).

For the weakly acidic GSE solutions subjected to heat treatment at 50 and 90 °C, the differences in absorbance and spectra were very small. At pH 6.7, increasing temperature from 50 to 90 °C obviously increased the absorbance at 280 nm, and a broad absorption band at about 500 nm appeared. Furthermore, GSE solution displayed higher absorption intensity and more shoulder bands under alkaline conditions than at pH 6.7 at two temperatures. These phenomena indicate that GSE has increased solubility under neutral and alkaline conditions, and at the same time, the polyphenolic compounds of GSE are not stable at high pHs and high temperatures. The instability of GSE stems from the oxidation reaction of polyphenolic compounds at high pHs, which creates new bonds and new structures [31,32]. The spectrophotometric study suggests that the application of GSE under weakly acidic conditions exhibits good stability.

3.2. Application Conditions of GSE

Four application conditions of GSE (pH, temperature, time, and GSE concentration) were discussed. Because the apparent color strength (K/S) cannot reflect the color depth of dyed fabrics accurately when the changes in color occur in some cases, both K/S and color difference (ΔE) were used as evaluation indexes for studying the color of cotton fabric in the present study.

3.2.1. pH

Considering the effect of pH on the stability of proanthocyanins, the pH of the GSE solution was first studied. Figure 3a displays that the K/S and ΔE values of cotton fabric were low at pH 6 and 7 whereas the two values were high at pH 4 and in the absence of pH regulator. Indeed, the pH of the GSE solution used in this section was 4.5 or so due to the ionization of phenolic hydroxyl groups in proanthocyanins, although pH regulator was not added. Therefore, the color depth at pH 4 was comparable with that in the absence of a pH regulator.

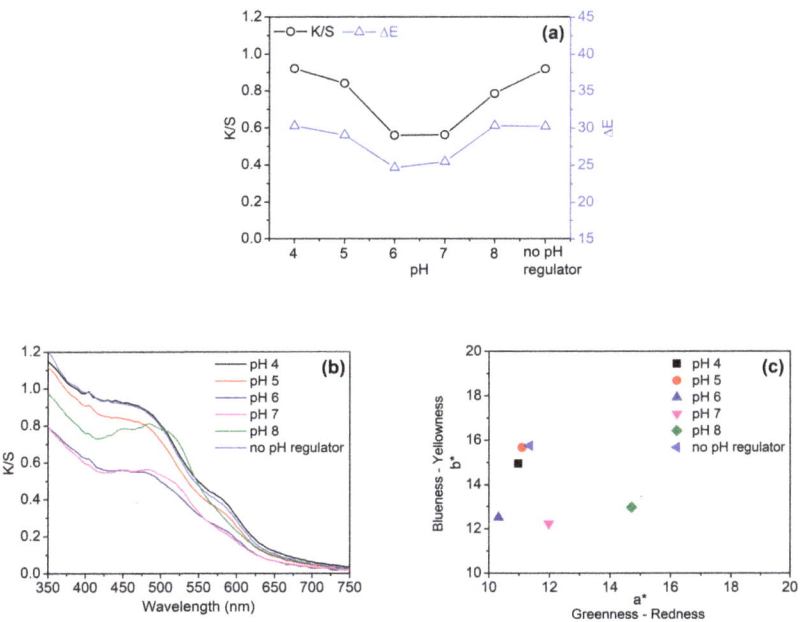

Figure 3. Color depth (K/S at 450 nm) and color difference (**a**), visible absorption spectra (**b**), and chromaticity coordinates (**c**) of cotton fabrics dyed with 20%owf GSE at various pH values.

The pH of the GSE solution had certain effects on the color hue of cotton fabric, as shown in Figure 3b,c. When pH was changed from 6 to 8, the maximum absorption wavelength of the dyed fabric shifted towards the long wavelength (Figure 3b). This red shift effect is caused by the ionization and oxidation of phenolic hydroxyl groups in proanthocyanins [32]. The changes in color at pH 7 and 8 were also reflected in the chromaticity coordinates of Figure 3c. The chromaticity coordinates of the fabrics dyed at these two pHs shifted more to the red color space. From Figure 3b,c, it can be observed that the visible absorption spectrum and chromaticity coordinate of the fabric dyed in the absence of pH regulator was very similar to those of the fabric dyed at pH 4. Taking the aforementioned results into consideration, a pH regulator was not used in the subsequent dyeing experiments.

3.2.2. Temperature and Time

Figure 4a,b shows the effects of temperature and time on the K/S and ΔE values of cotton fabrics, respectively. From Figure 4a, it can be observed that the color depth of cotton fabric increased gradually when temperature was elevated from 60 to 90 °C. This phenomenon can be explained by the increased swelling of cotton fiber and the increased diffusion kinetic energy of proanthocyanins at high temperatures. Overall, the increment of color depth caused by increasing temperature is not high, which is associated with the low affinity of GSE to cotton fiber indicated by low K/S and ΔE values. This result is not completely the same as that of silk fabric dyed with condensed tannins extracted from *Dioscorea cirrhosa* tuber [7]. For the dyeing of silk, the adsorption of condensed tannins increased significantly with increasing temperature due to the high affinity of condensed tannins to silk. Similarly, time had a limited effect on the color depth of cotton fabric, although a slight increase in color depth was observed with prolonged time (Figure 4b).

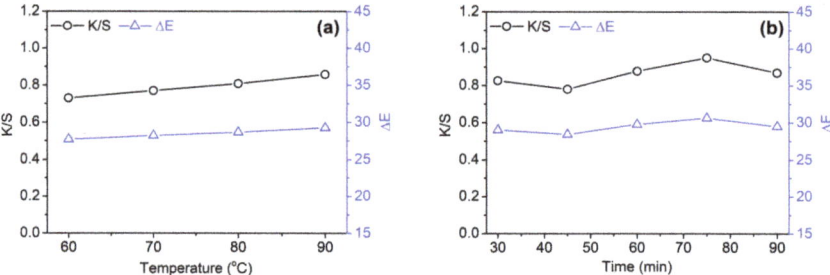

Figure 4. Effects of temperatures (**a**) and time (**b**) on the color depth (K/S at 450 nm) and color difference of cotton fabrics dyed with 20%owf GSE.

3.2.3. GSE Concentration

Figure 5a shows that the color depth of cotton fabric increased almost linearly with GSE concentration. Accordingly, the chromaticity coordinate moved upwards along a straight line (Figure 5b). This reflects the increase in chroma (C*), and also indicates the stability of color hue at different color depths.

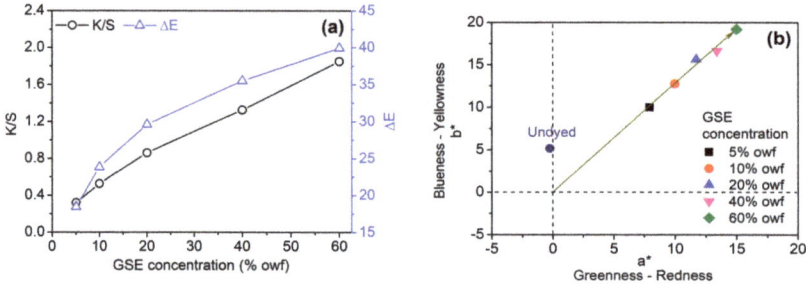

Figure 5. Color depth (K/S at 450 nm) and color difference (**a**), and chromaticity coordinates (**b**) of cotton fabrics dyed with GSE at various concentrations.

From the above experiments, it is clear that pH and GSE concentration have a great impact on the color depth of cotton fabric. No use of a pH regulator can impart the stable color to cotton fabric, and the color depth can be adjusted by the change of GSE concentration. In such dyeing conditions, the color hue of cotton fabric is easily controlled, and the resulting colors are in the light color category.

3.3. Color Fastness

The color fastness of cotton fabrics dyed with 10% and 20%owf GSE was determined and are listed in Table 2. In the grade classification of fastness standards, the highest grade of washing and rubbing fastness is 5. At 10% and 20%owf GSE, the dyed fabrics reached a high level in the fastness to washing and rubbing, as expected. The good fastness is associated with the fact that the colors in the present study belong to the light category. According to the Chinese national standard GB/T 18401–2010: National General Safety Technical Code for Textile Products, the acceptable color change fastness to washing must be higher than or equal to 3–4 for baby clothing, and 3 for the textile products in direct and indirect contact with skin [33]. For the dyed fabric with 20%owf GSE, the color change fastness to washing was 3–4. This grade can meet the aforementioned requirement for color fastness.

Table 2. Color fastness of cotton fabrics dyed with GSE.

Sample	Washing			Rubbing	
	Color Change	Staining		Dry	Wet
		Cotton	Wool		
10% GSE dyed	4–5	5	5	4–5	4–5
20% GSE dyed	3–4	4–5	5	4–5	4–5

3.4. Antibacterial and Antioxidant Properties

Textiles are often exposed to contamination with microbes during usage and storage. Cotton clothes are well known to have no antibacterial property. The microbial attack towards cotton textiles results in the changes in the color and appearance of fabrics, the strength loss of fabrics, unpleasant odor formation, and infectious disease. The most popular antibacterials used for textiles are metals and metal salts (e.g., silver), quaternary ammonium compounds, silane quaternary ammonium compounds, halogenated phenols (e.g., triclosan), metal-organic complexes, and polybiguanides [34]. Most of them have shortcomings of persistence in the environment, a potential for bioaccumulation, and dermal sensitization potential [34]. This is also the case for silane quaternary ammonium compounds that are extensively used for cotton products. In this regard, the eco-friendly products from plants can provide suitable alternatives [35].

Previous studies have demonstrated the inhibition effect of tannins from *Punica granatum*, *Quercus infectoria*, and *Dioscorea cirrhosa* towards the growth of microbes on textiles [7,8,11]. However, the antibacterial activity varied greatly with the different botanical sources and dosages of tannins [8,11]. In addition, lignins, which are similar to tannins in chemical structures, had bactericidal activity for eight bacteria cultures when applied onto linen fabric using a padding method [36]. At present, the antibacterial function of tannins from GSE for textile application remains unclear. In the present study, the inhibition effect of cotton fabrics dyed with 5% and 10%owf GSE towards the growth of *E. coli* was tested. Figure 6 shows distinct differences in the visual bacterial cultures between the undyed and dyed fabrics. For the fabric dyed with 10%owf GSE, bacterial colonies were hardly found. The sample had a high antibacterial rate of 96.3% in comparison to the undyed fabric, indicating excellent antibacterial function. In addition, the fabric dyed with 5%owf GSE displayed remarkably reduced bacterial colonies, and its antibacterial rate was 77.7%. This fabric can also be classified as an antibacterial textile according to the Chinese national standard GB/T 20944.3–2008 which requires the antibacterial products to have a bacteria inhibition rate of higher than 70% [27].

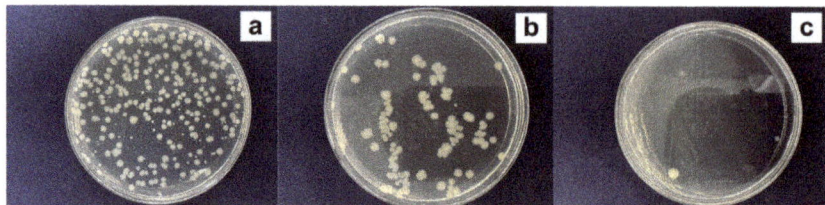

Figure 6. Visual bacterial cultures for undyed cotton fabric (**a**) and dyed cotton fabrics with 5% (**b**) and 10%owf (**c**) GSE.

The antioxidant activity of textiles has not attracted enough attention in the past. Indeed, clothes with an antioxidant function can protect the skin against the free radicals present in the atmosphere, which are responsible for skin aging. Antioxidants can scavenge free radicals, thus deactivating the capacity of free radicals to damage the skin when incorporated into textile fibers. It is well known that the extracts from grape seeds and skins have high antioxidant capability [16,18,37]. The antioxidant property of GSE is attributed to the action of polyphenols and proanthocyanidins [37].

Herein, the antioxidant activity of cotton fabrics dyed with GSE at various concentrations was evaluated using the $ABTS^+$ decolorization assay. In this assay, the fabrics were soaked into the $ABTS^+$ solution to scavenge $ABTS^+$, leading to the decolorization of $ABTS^+$. Thus, a high decolorization rate of $ABTS^+$ represents a high antioxidant activity. Figure 7 shows that pristine cotton had no antioxidant function (decolorization rate, 6.3%). When GSE concentration increased from 5% to 20%owf, the decolorization rate of $ABTS^+$ increased from 42.4% to 97.1%. At 20%owf GSE, the decolorization rate almost reached a plateau. This test proves the high efficiency of GSE on cotton fabric in scavenging free radicals.

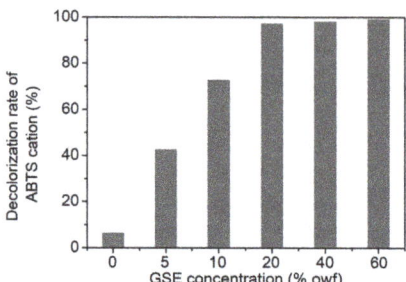

Figure 7. Antioxidant activity of cotton fabrics dyed with GSE at various concentrations.

3.5. UV Protection Ability

Cotton garments, which are often used in summer, cannot provide sufficient protection against solar radiation. Thus the treatment with synthesized UV absorbers containing reactive groups must be carried out [38,39]. Some reports have demonstrated that natural dyes can also impart UV protection effects to cotton textiles [40,41]. As natural compounds structurally similar to tannins, nanolignins have also been proven to be excellent UV absorbing agents for increasing the UV protection of linen fabric [36]. Such function is attributable to the strong absorption of lignins in the UV light region [42].

Figure 8a displays the UV transmission curves of cotton fabrics. The pristine fabric showed a high transmittance in the region of both the UVB (280–315 nm) and UVA (315–400 nm), with a low UPF (11.75). The fabrics dyed with GSE showed the evident reduction in the transmittance of UVA and UVB, as indicated by Figure 8a,b. Moreover, the UV transmittance decreased and the UPF increased with increasing GSE concentration. The UPF of the fabrics dyed with 5% and 10%owf GSE was 49.56 and 73.53, respectively. Correspondingly, the UVA transmittance (T_{UVA}) was 5.19% and 3.35%, respectively, whereas the UVB transmittance (T_{UVB}) was 1.56% and 1.06%, respectively. In respect to the Australia/New Zealand standard AS/NZS 4399:1996 [43], the UPF rating of the two fabrics reached 40–50 and 50+, respectively, both of which are classified as "excellent protection". The excellent UV protection capability of the GSE dyed fabric results from a number of aromatic rings present in the proanthocyanins of GSE.

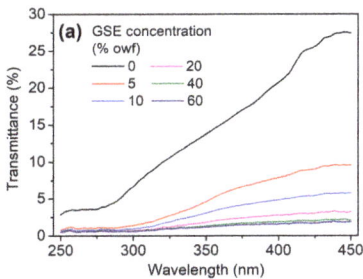 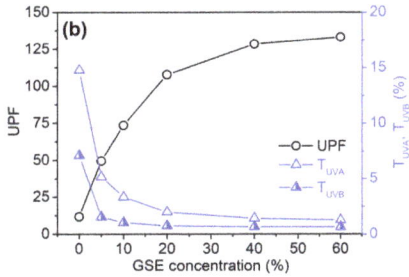

Figure 8. UV transmittance curves (**a**) as well as UPF and UVA/UVB transmittance (**b**) of cotton fabrics dyed with GSE at various concentrations.

4. Conclusions

The present study discusses the conditions of cotton dyeing with GSE as well as the color stability and color fastness of the dyed fabric and evaluates the antibacterial, antioxidant and UV protection functions of the dyed fabric. The spectrophotometric study demonstrated the good stability of GSE in a weakly acidic medium. In the case of no use of pH regulator, the color hue of dyed fabric was stable because of the weakly acidic nature of the GSE solution. Without the use of a pH regulator, the color depth of the dyed fabric was mainly dependant on GSE concentration and less affected by temperature and time. The functional properties of the dyed fabric depended on the GSE concentration. Excellent antibacterial, antioxidant, and UV protection functions were achieved at 10%, 20%, and 5%owf GSE, respectively. The present study is the preliminary exploration of the application of GSE in the simultaneous dyeing and functional modification of cotton fabric. In this study, pale colors were obtained. Further studies should be performed to increase the color depth, color fastness, and washing durability of the GSE-dyed cotton fabric.

Author Contributions: Conceptualization, L.G. and R.-C.T.; methodology, L.G. and Z.-Y.Y.; formal analysis and investigation, L.G., Z.-Y.Y. and H.-B.Y.; data curation, L.G. and H.-B.Y.; writing—original draft preparation, L.G.; writing—review and editing, R.-C.T.; and supervision, R.-C.T. All authors have read and agreed to the published version of the manuscript.

Funding: This research was funded by the Priority Academic Program Development (PAPD) of Jiangsu Higher Education Institutions [PAPD-2018-87].

Conflicts of Interest: The authors declare no conflict of interest.

Abbreviations

The following abbreviations are used in this manuscript:

a*	Redness-greenness index
ABTS	2,2′-Azino-bis (3-ethylbenzothiazoline-6-sulphonic acid) diammonium salt
AS/NZS	Australia/New Zealand standard
b*	Yellowness-blueness index
C*	Chroma
E. coli	*Escherichia coli*
ΔE	Color difference
GB/T	Chinese national standard suggested

GSE	Grape seed extract
ISO	International Organization for Standardization
K/S	Apparent color strength
L*	Lightness
owf	On the weight of fabric
UPF	Ultraviolet protection factor
UV	Ultraviolet
UVA	Ultraviolet light from 315 nm to 400 nm
UVB	Ultraviolet light from 280 nm to 315 nm
UV–Vis	Ultraviolet–visible

References

1. Arbenz, A.; Avérous, L. Chemical modification of tannins to elaborate aromatic biobased macromolecular architectures. *Green Chem.* **2015**, *17*, 2626–2646. [CrossRef]
2. Duval, A.; Couture, G.; Caillol, S.; Avérous, L. Biobased and aromatic reversible thermoset networks from condensed tannins via the Diels–Alder reaction. *ACS Sustain. Chem. Eng.* **2017**, *5*, 1199–1207. [CrossRef]
3. Pizzi, A. Tannins: prospectives and actual industrial applications. *Biomolecules* **2019**, *9*, 344. [CrossRef] [PubMed]
4. Blanchart, P.; Dembelé, A.; Dembelé, C.; Pléa, M.; Bergström, L.; Granet, R.; Sol, V.; Gloaguen, V.; Degot, M.; Krausz, P. Mechanism of traditional Bogolan dyeing technique with clay on cotton fabric. *Appl. Clay Sci.* **2010**, *50*, 455–460. [CrossRef]
5. Restivo, A.; Degano, I.; Ribechini, E.; Pérez-Arantegui, J.; Colombini, M.P. Field-emission scanning electron microscopy and energy-dispersive X-ray analysis to understand the role of tannin-based dyes in the degradation of historical wool textiles. *Microsc. Microanal.* **2014**, *20*, 1534–1543. [CrossRef]
6. Wilson, H.; Carr, C.; Hacke, M. Production and validation of model iron-tannate dyed textiles for use as historic textile substitutes in stabilization treatment studies. *Chem. Cent. J.* **2012**, *6*, 44. [CrossRef]
7. Yang, T.-T.; Guan, J.-P.; Tang, R.-C.; Chen, G. Condensed tannin from *Dioscorea cirrhosa* tuber as an eco-friendly and durable flame retardant for silk textile. *Ind. Crops Prod.* **2018**, *115*, 16–25. [CrossRef]
8. Gupta, D.; Khare, S.K.; Laha, A. Antimicrobial properties of natural dyes against Gram-negative bacteria. *Color Technol.* **2004**, *120*, 167–171. [CrossRef]
9. Pisitsak, P.; Hutakamol, J.; Thongcharoen, R.; Phokaew, P.; Kanjanawan, K.; Saksaeng, N. Improving the dyeability of cotton with tannin-rich natural dye through pretreatment with whey protein isolate. *Ind. Crops Prod.* **2016**, *79*, 47–56. [CrossRef]
10. Rather, L.J.; Akhter, S.; Padder, R.A.; Hassan, Q.P.; Hussain, M.; Khan, M.A.; Mohammad, F. Colorful and semi durable antioxidant finish of woolen yarn with tannin rich extract of *Acacia nilotica* natural dye. *Dyes Pigm.* **2017**, *139*, 812–819. [CrossRef]
11. Singh, R.; Jain, A.; Panwar, S.; Gupta, D.; Khare, S.K. Antimicrobial activity of some natural dyes. *Dyes Pigm.* **2005**, *66*, 99–102. [CrossRef]
12. Medouni-Adrar, S.; Boulekbache-Makhlouf, L.; Cadot, Y.; Medouni-Haroune, L.; Dahmoune, F.; Makhoukhe, A.; Madani, K. Optimization of the recovery of phenolic compounds from Algerian grape by-products. *Ind. Crops Prod.* **2015**, *77*, 123–132. [CrossRef]
13. Drosou, C.; Kyriakopoulou, K.; Bimpilas, A.; Tsimogiannis, D.; Krokida, M. A comparative study on different extraction techniques to recover red grape pomace polyphenols from vinification byproducts. *Ind. Crops Prod.* **2015**, *75*, 141–149. [CrossRef]
14. Dwyer, K.; Hosseinian, F.; Rod, M. The market potential of grape waste alternatives. *J. Food Res.* **2014**, *3*, 91–106. [CrossRef]
15. Tounsi, M.S.; Ouerghemmi, I.; Wannes, W.A.; Ksouri, R.; Zemni, H.; Marzouk, B.; Kchouk, M.E. Valorization of three varieties of grape. *Ind. Crops Prod.* **2009**, *30*, 292–296. [CrossRef]
16. Mandic, A.I.; Đilas, S.M.; Ćetković, G.S.; Čanadanović-Brunet, J.M.; Tumbas, V.T. Polyphenolic composition and antioxidant activities of grape seed extract. *Int. J. Food Prop.* **2008**, *11*, 713–726. [CrossRef]
17. Sarni-Manchado, P.; Cheynier, V.; Moutounet, M. Interactions of grape seed tannins with salivary proteins. *J. Agric. Food Chem.* **1999**, *47*, 42–47. [CrossRef]

18. Jayaprakasha, G.K.; Selvi, T.; Sakariah, K.K. Antibacterial and antioxidant activities of grape (*Vitis vinifera*) seed extracts. *Food Res. Int.* **2003**, *36*, 117–122. [CrossRef]
19. Shankar, S.; Rhim, J.-W. Antimicrobial wrapping papercoated with a ternary blend of carbohydrates (alginate, carboxymethyl cellulose, carrageenan) and grapefruit seed extract. *Carbohyd. Polym.* **2018**, *196*, 92–101. [CrossRef]
20. Xia, E.-Q.; Deng, G.-F.; Guo, Y.-J.; Li, H.-B. Biological activities of polyphenols from grapes. *Int. J. Mol. Sci.* **2010**, *11*, 622–646. [CrossRef]
21. Baaka, N.; Haddar, W.; Ben Ticha, M.; Amorim, M.T.P.; M'Henni, M.F. Sustainability issues of ultrasonic wool dyeing with grape pomace colorant. *Nat. Prod. Res.* **2017**, *31*, 1655–1662. [CrossRef] [PubMed]
22. Baaka, N.; Ben Ticha, M.; Haddar, W.; Mhenni, M.F. A challenging approach to improve the low affinity of acrylic fibers to be successfully dyed with a bio-colorant extracted from grape marc. *J. Renew. Mater.* **2019**, *7*, 289–300. [CrossRef]
23. Yang, H.; Park, Y. Optimum dyeing condition of cotton by fermented grape by-products with degraded protein mordant. *Text. Coloration Finish.* **2015**, *27*, 202–209. [CrossRef]
24. Prieur, C.; Rigaud, J.; Cheynier, V.; Moutounet, M. Oligomeric and polymeric procyanidins from grape seeds. *Phytochemistry* **1994**, *36*, 781–784. [CrossRef]
25. Haslam, E. Natural polyphenols (vegetable tannins) as drugs: possible modes of action. *J. Nat. Prod.* **1996**, *59*, 205–215. [CrossRef]
26. Rigg, B. Chapter 3 Colorimetry and the CIE system. In *Color Physics for Industry*, 2nd ed.; McDonald, R., Ed.; Society of Dyers and Colorists: West Yorkshire, UK, 1997; pp. 81–120.
27. *Textiles–Evaluation for Antibacterial Activity–Part 3: Shake Flask Method.* GB/T 20944.3–2008, China's General Administration of Quality Supervision; Inspection and Quarantine and Standardization Administration of China: Beijing, China, 2008.
28. Re, R.; Pellegrini, N.; Proteggente, A.; Pannala, A.; Yang, M.; Rice-Evans, C. Antioxidant activity applying an improved ABTS radical cation decolorization assay. *Free Radic. Biol. Med.* **1999**, *26*, 1231–1237. [CrossRef]
29. Zhang, W.; Yang, Z.-Y.; Cheng, X.-W.; Tang, R.-C.; Qiao, Y.-F. Adsorption, antibacterial and antioxidant properties of tannic acid on silk fiber. *Polymers* **2019**, *11*, 970. [CrossRef]
30. Falcão, L.; Araújo, M.E.M. Tannins characterization in historic leathers by complementary analytical techniques ATR-FTIR, UV-Vis and chemicaltests. *J. Cult. Herit.* **2013**, *14*, 499–508. [CrossRef]
31. Makkar, H.P.S.; Becker, K. Effect of pH, temperature, and time on inactivation of tannins and possible implications in detannification studies. *J. Agric. Food Chem.* **1996**, *44*, 1291–1295. [CrossRef]
32. Friedman, M.; Jürgens, H.S. Effect of pH on the stability of plant phenolic compounds. *J. Agric. Food Chem.* **2000**, *48*, 2101–2110. [CrossRef] [PubMed]
33. *National General Safety Technical Code for Textile Products.* GB/T 18401–2010, China's General Administration of Quality Supervision; Inspection and Quarantine and Standardization Administration of China: Beijing, China, 2010.
34. Windler, L.; Height, M.; Nowack, B. Comparative evaluation of antimicrobials for textile applications. *Environ. Int.* **2013**, *53*, 62–73. [CrossRef] [PubMed]
35. Shahid-ul-Islam; Mohammad, F. Natural colorants in the presence of anchors so-called mordants as promising coloring and antimicrobial agents for textile materials. *ACS Sustain. Chem. Eng.* **2015**, *3*, 2361–2375. [CrossRef]
36. Zimniewska, M.; Kozłowski, R.; Batog, J. Nanolignin modified linen fabric as a multifunctional product. *Mol. Cryst. Liq. Cryst.* **2008**, *484*. [CrossRef]
37. GöktürkBaydar, N.; Özkan, G.; Yaşar, S. Evaluation of the antiradical and antioxidant potential of grape extracts. *Food Control* **2007**, *18*, 1131–1136. [CrossRef]
38. Gorjanc, M.; Jazbec, K.; Mozetič, M.; Kert, M. UV protective properties of cotton fabric treated with plasma, UV absorber, and reactive dye. *Fiber. Polym.* **2014**, *15*, 2095–2104. [CrossRef]
39. Sahar, A.; Ali, S.; Hussain, T.; Irfan, M.; Eliasson, B.; Iqbal, J. UV absorbers for cellulosic apparels: A computational and experimental study. *Spectrochim. Acta A* **2018**, *188*, 355–361. [CrossRef]
40. Grifoni, D.; Bacci, L.; Di Lonardo, S.; Pinelli, P.; Scardigli, A.; Camilli, F.; Sabatini, F.; Zipoli, G.; Romani, A. UV protective properties of cotton and flax fabrics dyed with multifunctional plant extracts. *Dyes Pigm.* **2014**, *105*, 89–96. [CrossRef]
41. Shahid, M.; Shahid-ul-Islam; Mohammad, F. Recent advancements in natural dye applications: A review. *J. Clean. Prod.* **2013**, *53*, 310–331. [CrossRef]

42. Mishra, P.K.; Wimmer, R. Aerosol assisted self-assembly as a route to synthesize solid and hollowspherical lignin colloids and its utilization in layer by layer deposition. *Ultrason. Sonochem.* **2017**, *35*, 45–50. [CrossRef]
43. *Sun Protective Clothing—Evaluation and Classification*; Homebush NSW: Wellington, New Zealand, 1996.

© 2020 by the authors. Licensee MDPI, Basel, Switzerland. This article is an open access article distributed under the terms and conditions of the Creative Commons Attribution (CC BY) license (http://creativecommons.org/licenses/by/4.0/).

Article

The Effect of the Addition of Blue Honeysuckle Berry Juice to Apple Juice on the Selected Quality Characteristics, Anthocyanin Stability, and Antioxidant Properties

Anna Grobelna [1], Stanisław Kalisz [1,*] and Marek Kieliszek [2,*]

1. Department of Food Technology and Assessment, Institute of Food Sciences, Warsaw University of Life Sciences—SGGW, Nowoursynowska 159 C, 02-776 Warsaw, Poland; anna_grobelna@sggw.pl
2. Department of Food Biotechnology and Microbiology, Institute of Food Sciences, Warsaw University of Life Sciences—SGGW, Nowoursynowska 159 C, 02-776 Warsaw, Poland
* Correspondence: stanislaw_kalisz@sggw.pl (S.K.); marek_kieliszek@sggw.pl (M.K.)

Received: 1 November 2019; Accepted: 14 November 2019; Published: 17 November 2019

Abstract: Apple juice is rich in phenolic compounds that are important as natural antioxidants. In turn, blue honeysuckle berry juice is a valuable source of bioactive ingredients and can be an interesting and beneficial supplement to fruit juices. The aim of this study was to examine the physicochemical and sensory properties of the newly designed mixture of apple juice and blue honeysuckle berry juice. The addition of blue honeysuckle berry juice to apple juice had a significant effect on the content of anthocyanin and vitamin C in the newly designed fruit juices. After production, the content of anthocyanins and polyphenols in the blue honeysuckle berry juice was high (595.39 and 767.88 mg/100 mL, respectively). As the concentration of blue honeysuckle berry juice added to apple juice was increased, the polyphenol content also increased. The juices analyzed after 4 months of storage were lighter and showed a less intense red color than the juices analyzed directly after production. Antioxidant activity (ABTS assay) in the apple juice mixed with 10% blueberry juice was almost 3 times higher than the pure apple juice after 3 months of storage; the addition of 30% blueberry juice significantly increased the antioxidant activity of the apple juice. Thus, the results of this research have expanded the existing knowledge about the health and sensory properties of apple juice mixed with blue honeysuckle berry juice. These findings can be utilized in further research aiming at the development of new products that can meet consumer expectations.

Keywords: blue honeysuckle berry; apple; anthocyanins; polyphenols; antioxidant; juice

1. Introduction

Fruit juices are important products for consumers looking for an alternative to fresh fruit. These are widely consumed by most people because of their freshness, sensory properties, and nutritional value. Increased expectations and increased consumer awareness of the modern methods of processing fruits and vegetables have resulted in the continuous development of the fresh juice industry. The contemporary food industry could not be developed without the production of new products that can exhibit health-promoting effects in addition to reducing the risk of certain diseases. Consumers value those raw materials that are known and liked by them, but are increasingly aware of the impact on health by foods produced using new and less used raw materials.

Apple juice is one of the most popular products in the world, both in terms of the production process and international exchange. Apples play a very important role in the food industry and are one of the fruits widely grown around the world. The largest apple producers in the world include, among others, China, the United States of America (USA), Poland, and Turkey [1]. Currently, several thousand varieties of apples are grown. Old, almost forgotten varieties of apples, which are characterized by a juicy, delicate, and sweet and sour pulp, are becoming more popular. Apple has a pleasant taste and is rich in the gelling agent pectin, which prevents the phase separation of juices [2]. These fruits are also rich in nutrients that have a positive effect on the human body. In addition, apples are a natural and rich source of compounds with antioxidant properties (phenols). However, it should be noted that the content of phenolic compounds in apples is highly dependent on their variety and cultivation practices [3]. These fruits also possess minerals (potassium, magnesium, iron, calcium), quercetin, luteolin, apigenin, ursolic acid, and fiber [4,5]. Moreover, apple juices attract consumers with their sensory properties and can thus be an important part of a menu. It is worth emphasizing that the high-quality apple juice concentrate produced in Poland is more popular among foreign customers. The main advantage of this concentrate is its acidity, the level of which is determined by the variety and climate during the production of fruits.

In the era of modern globalization, the pursuit of design and production of new food products and their usefulness in processing have resulted in an increase in consumer interest in mixed juices obtained from various fruits. Growing production and consumption of juices incline us to look for ways to make them more attractive sensorially and nutritionally by including other raw materials, with blue honeysuckle berry (*Lonicera caerulea* L.) being one such component. Blue honeysuckle berry is also commonly known as scotch, blue scrub, or haskap. *Lonicera caerulea* L. is a perennial fruit plant belonging to the family Caprifoliaceae (Figure 1) [6,7].

Figure 1. Fruit of blue honeysuckle berry.

Its fruits are a valuable, natural source of vitamin C (in the level ranging from 29 to 187 mg/100 g) and anthocyanins (including cyanidin-3-*O*-glucoside). B-group vitamins are also found in smaller amounts in their berries [8]. Due to the beneficial chemical composition and attractive aroma, these berries have been used in the food-processing industry as a valuable addition to juices and purees. Undoubtedly, blue honeysuckle berries have greater health-promoting properties than the other commonly consumed berries. The presence of anthocyanins in blue honeysuckle berries contributes to their antioxidant effects [9]. It is believed that blue honeysuckle berry was used in folk medicine to reduce the risk of hypertension, glaucoma, anemia, osteoporosis [10,11], and gastrointestinal disorders [6], and in the treatment of various eye diseases [8]. The current scientific literature has only scarce data on the enrichment of apple juice with blue honeysuckle berry juice, and hence, it would be interesting to highlight the progress of work in the juice-processing area of the food industry. The growing interest of consumers in fresh beverages with high nutritional value, which are healthy and ready to drink with satisfactory organoleptic properties, encourages producers to introduce new products. This systematic support to develop novel products of consumer and scientific interests has led us to combine two products, one having an established position in the fruit and vegetable market (apple) and the other (raw material) that is currently becoming popular (blue honeysuckle berry). To our knowledge, this is the first report on the physicochemical characteristics of combined apple and berry juices.

The aim of this study was to assess the chemical and sensory properties of mixed apple and blue honeysuckle berry juice. The content of polyphenols in the mixtures of apple juice and blue honeysuckle berry juice was evaluated after a storage period, and a consumer analysis was carried out. In addition, a detailed analysis of the physicochemical parameters (including antioxidant activity, content of anthocyanins, content of vitamin C) was performed.

2. Materials and Methods

2.1. Reagents and Standards

Acetonitrile, formic acid, phosphoric acid, 2,2′-azino-bis (3-ethylbenzothiazoline-6-sulfonic acid) (ABTS), 6-hydroxy-2,5,7,8-tetramethylchroman-2-carboxylic acid (Trolox), Folin–Ciocalteu reagent, and sodium carbonate were purchased from Sigma-Aldrich (Steinheim, Germany). Cyanidin-3,5-O-diglucoside, cyanidin-3-O-glucoside, and cyanidin-3-O-rutinoside, peonidin-3-O-rutinoside and peonidin-3-O-glucoside, pelargonidin-3-O-glucoside, ascorbic acid, and gallic acid were purchased from Extrasynthese (Lyon, France).

2.2. Raw Materials

Blue honeysuckle berries (*L. caerulea* L. cv. Dlinnoplodna) were obtained from the experimental orchard of the Research Institute of Horticulture in Skierniewice, Poland. Apples (*Malus domestica* Borkh. cv. Champion) were obtained from the experimental field of the Department of Pomology of the Warsaw University of Life Sciences, Poland. The blue honeysuckle fruits were harvested in the summer of 2017 and then frozen and stored at −29 °C until the juice was pressed, while the apples were harvested in the autumn of 2017 and then stored at 4 °C until juice extraction.

2.3. The Technology of Juice Production

Prior to juice pressing, the raw materials were pretreated. Briefly, the apples were washed and sliced, while the blue honeysuckle berries were thawed at 25 °C. Afterward, the blue honeysuckle berry pulp and apple pulp, at an amount of 0.44 and 1 g/kg, respectively, were enzymatically treated with Rohapect 10 L (AB Enzymes GmbH, Germany). The raw materials thus prepared were pressed using a laboratory press. The obtained apple juice and the blue honeysuckle berry juice were mixed in the proportions of 90:10, 80:20, and 70:30, respectively. Altogether, five different juice variants were produced (Table 1).

Table 1. The types of juices obtained.

Symbol of the Juice Type	The Percentage Share of the Individual Juice [%]
A	100% A
H	100% H
AH1	90% A-10% H
AH2	80% A-20% H
AH3	70% A-30% H

A Juice from apple; H Juice from blue honeysuckle berry.

The obtained juices were poured into glasses and packed, following which they were pasteurized at 85 °C for 15 min and immediately cooled to 20 °C. The juices were stored at 20 °C, in the absence of light, in order to limit the effect of external factors on the content of the biologically active compounds. All of the obtained juices were analyzed immediately after production and also after 1, 2, 3, and 4 months of storage at 20 °C. The analyses were performed in three replicates.

2.4. Analytical Methods

2.4.1. Physicochemical Parameters

The content of total soluble solids (TSS) was measured with the automatic refractometer Refracto 30PX (Mettler Toledo, Poland). For this purpose, a few drops of juice were applied to the prism of the device, and the reading was taken with an accuracy of 0.1. The active acidity (pH) was determined using the pH meter Hi 221 (Hanna Instruments, Poland), which was calibrated using the buffers of pH 4 and 7. The measurement was read from the display of the electronic pH meter with an accuracy of 0.01. To determine the titratable acidity (TTA) of the juice samples, a potentiometric titration with 0.1 M NaOH solution was performed using the pH meter Hi 221 (Hanna Instruments) until a pH of 8.1 was reached. Acidity was expressed in g of malic acid/100 mL of juice, as described by Wojdyło et al. [12]

2.4.2. HPLC Analysis of Anthocyanins

The content of anthocyanins in the tested juices was determined using high-performance liquid chromatography (HPLC) in an isocratic system with a Luna column (5 µm, C18(2), 250 × 4.6 mm, Phenomenex), as described by Goiffon et al. [13]. The flow rate was fixed as 1 mL/min, and the temperature was set at 25 °C. The mobile phase consisted of a mixture of water, acetonitrile, and formic acid at a volume ratio of 810:90:100. Before the analysis, the juices were passed through PTFE syringe filters with a pore size of 0.45 µm. The results were recorded at a wavelength (λ) of 520 nm. The total anthocyanin content was expressed in mg/100 mL of juice.

2.4.3. HPLC Analysis of L-Ascorbic Acid

The content of L-ascorbic acid in the tested juices was determined by applying the HPLC method described by Oszmiański and Wojdyło [14] using an Onyx Monolithic C18 column (100 × 4.6 mm, Phenomenex). The eluent used was 0.1% solution of H_3PO_4. Before the analysis, the juices were passed through PTFE syringe filters with a pore size of 0.45 µm. The results were recorded at λ = 254 nm, and the content of L-ascorbic acid was expressed in mg/100 mL of juice.

2.4.4. Analysis of Total Phenolic Content

The determination of total polyphenols (TP) was performed by the method with Folin–Ciocalteu reagent according to Gao et al. [15]. From the calibration curve, the quantities of milligrams of gallic acid corresponding to the investigated absorbance values were calculated. The results are shown in milligrams of gallic acid per 100 mL of juice.

2.4.5. Antioxidant Activity ($ABTS^+$) Assay

The antioxidant activity was determined in the juices according to the method of Re et al. [16]. Briefly, 40 µL of juice was taken in tubes. Then, 4 mL of $ABTS^+$ cation radical solution was added and the tubes were stirred. Six minutes after the addition of the cation radical solution, the juice samples were taken in 1 cm cuvettes and their absorbance was measured in relation to distilled water at 734 nm using a spectrophotometer (Shimadzu UV—1650PC). The antioxidant activity of the tested extracts was expressed in µmol Trolox/mL [17].

2.4.6. Juice Color Parameters

The color of the juices was analyzed with the Konica Minolta CM-3600d colorimeter equipped with SpectraMagic NX program. Optical cuvettes with an optical path length of 2 mm, 10-degree observer, and D65 illuminant were used for the measurement. The results are presented in the CIE system L*a*b* with an accuracy of 0.01, where the parameter a* describes the share of green or red, parameter b* describes the share of yellow or blue, and the parameter L* corresponds to brightness.

On the basis of the measured color parameters (L*, a*, and b*), the total change in color (ΔE) was calculated using the formula Δ = as described by Wojdyło et al. [12].

2.5. Sensory Assessment

The sensory evaluation was carried out according to the method described by Lachowicz and Oszmiański [18]. A 5-point hedonic scale was used for evaluation, where 1 indicated fully unacceptable and 5 indicated fully acceptable. The evaluation was carried out by a group of 15 trained panelists. Samples of juices at 20 °C were provided in plastic and transparent cups for evaluation. Parameters such as taste, aroma, color, and general characteristics were assessed.

2.6. Statistical Analysis

Statistical analysis of the results was performed using Statistica (version 13.3) and Excel 2016. The standard deviation was calculated using Excel, while Statistica was used to perform analysis of variance (ANOVA) and to analyze the significance of differences using Tukey's test.

3. Results and Discussion

3.1. Physicochemical Parameters and Sensory Assessment

TSS value is a quality-control indicator commonly used in the juice industry [18]. The TSS values of juices measured immediately after production ranged from 13.54 (both A and AH1 variants) to 14.15 (H variant) (Table 2). It was found that the storage time had no significant effect on TSS values in any of the variants. The highest TSS values were recorded for the H variant (above 14.05 °Brix). The TSS value seemed to be affected by the degree of fragmentation of fruit tissues during enzymatic treatment. According to Nath et al. [19], a higher degree of tissue breakdown may contribute to a higher TSS value due to a higher release of compounds such as sugars. Compared to apples, blue honeysuckle berries are fleshy and, hence, yield easier to the crushing process.

The TTA and pH are the next important physicochemical parameters that determine the juice quality. They can significantly influence the stability of bioactive components present in fruit juices [11,20]. The results show significant differences in the TTA and pH values between the tested juices. The TTA values measured in juices immediately after production ranged from 0.56 (A juice) to 3.64 g/100 mL (H juice) (Table 2), while the pH values ranged from 2.63 (H juice) to 3.13 (A juice). The addition of blue honeysuckle berry juice in any proportion caused an increase in acidity and a decrease in pH in the apple juice. The higher the proportion of blue honeysuckle berry juice, the higher the TTA value and the lower the pH of the mixed juices (AH1, AH2, AH3).

The TSS/TTA ratio is a parameter that greatly influences the sensory acceptability and consumer preferences. The higher the TSS/TTA ratio of the juice, the sweeter and more acceptable it is by consumers [21]. In this study, it was found that immediately after production, blue honeysuckle berry juice was characterized by an exceptionally low TSS/TTA ratio (3.88), which indicated its lower suitability for direct consumption. By contrast, the apple juice had a higher TSS/TTA ratio (24.28), which justified its high acceptability and popularity among consumers [22]. The addition of the blue honeysuckle berry juice (H variant) to apple juice (A variant) in different proportions caused a decrease in the TSS/TTA ratio. The higher the proportion of H juice added, the lower the TSS/TTA ratio of the mixed juices.

To verify the sensory acceptability of the prepared juices, a sensory evaluation was performed using a 5-point hedonic scale (Figure 2). With respect to the aroma, the highest score was given to the A juice (4.9), while the H juice received the lowest score (3.2).

In terms of color, the A juice received the highest score (4.7) and the AH1 juice received the lowest score (4.3). The color of the obtained mixed juices was directly proportional to the percentage of H juice added. Visually, the color of the AH1 juice was the most light compared to the other mixed juices and the most different from the base 100% blue honeysuckle berry juice (H), which is also indicated by the results of parameter measurements colors using the CieLab method, where the AH1 juice showed the highest value of the L* parameter compared to other mixed juices and H juice. With respect to taste, the A juice achieved the highest score (4.7), which was probably due to the lowest acidity value, higher pH, highest TSS/TTA ratio, and the fact that it is the commonly known fruit juice and widely accepted by consumers [22], whereas the H juice received the lowest score (3.0), which was probably due to its very high acidity. Both the AH1 and AH3 juice variants received relatively high scores in terms of taste (4.5 and 4.4, respectively), which indicates that the mixed juices with 10% and 30% proportions of the blue honeysuckle berry juice would be acceptable by consumers. However, as regards taste, the variant with 20% of blue honeysuckle berry juice (AH2) received the lowest score (3.4), which confirms the fact that the percentages of individual juices should be optimally selected for the production of mixed juices. According to Lesschaeve and Noble [23], the sensory characteristics of beverages can be influenced by factors such as polyphenolic composition, pH, and sugar content. However, despite a lower pH and higher TSS/TTA ratio compared to the AH3 juice, the AH2 juice was less sensorially acceptable, mainly in terms of taste and aroma. In this study, fruit juices with different chemical compositions and polyphenol contents [8,22] were mixed, which could have led to interactions between individual components influencing the sensory features.

Table 2. Physicochemical parameters and L-ascorbic acid content in the tested juices after production and during 4 months of storage.

Juice	Parameters	Time of Storage				
		After Production	1 Month	2 Months	3 Months	4 Months
A	TSS [A]	13.54 ± 0.08 [a]	13.44 ± 0.00 [a]	13.50 ± 0.00 [a]	13.44 ±0.08 [a]	13.54 ± 0.08 [a]
	TTA [B]	0.56 ± 0.00 [a]	0.54 ± 0.00 [b]	0.50 ± 0.00 [c]	0.49 ± 0.00 [c]	0.50 ± 0.00 [c]
	TSS/TTA	24.17	24.88	27.00	27.42	27.08
	pH	3.13 ± 0.01 [c]	3.19 ± 0.01 [b]	3.24 ± 0.02 [b]	3.19 ± 0.02 [b]	3.51 ± 0.02 [a]
	L-ascorbic acid [C]	0.52 ± 0.00 [a]	0.44 ± 0.00 [b]	0.31 ± 0.00 [c]	0.24 ± 0.00 [d]	0.17 ± 0.01 [e]
H	TSS [A]	14.15 ± 0.08 [a,b]	14.05 ± 0.08 [b]	14.25 ± 0.08 [a]	14.15 ±0.08 [a,b]	14.12 ± 0.02 [a,b]
	TTA [B]	3.64 ± 0.02 [a]	3.63 ± 0.02 [a]	3.59 ± 0.02 [a,b]	3.56 ± 0.02 [b]	3.56 ± 0.02 [b]
	TSS/TTA	3.88	3.87	3.96	3.97	3.96
	pH	2.63 ± 0.00 [c]	2.67 ± 0.00 [b]	2.68 ± 0.00 [b]	2.68 ± 0.01 [b]	2.72 ± 0.00 [a]
	L-ascorbic acid [C]	32.59 ± 0.19 [a]	11.98 ± 0.07 [b]	11.62 ± 0.07 [c]	10.44 ± 0.06 [d]	9.36 ± 0.05 [e]
AH1	TSS [A]	13.54 ± 0.08 [a]	13.53 ± 0.08 [a]	13.54 ± 0.08 [a]	13.54 ± 0.08 [a]	13.64 ± 0.08 [a]
	TTA	0.78 ± 0.01 [a]	0.76 ± 0.00 [b]	0.74 ± 0.01 [c]	0.76 ± 0.00 [b]	0.69 ± 0.00 [d]
	TSS/TTA	17.35	17.80	18.29	17.81	19.76
	pH	3.02 ± 0.01 [c]	3.08 ± 0.00 [b]	3.08 ± 0.00 [b]	3.08 ± 0.02 [b]	3.40 ± 0.01 [a]
	L-ascorbic acid [C]	1.14 ± 0.01 [a]	0.86 ± 0.00 [b]	0.52 ± 0.00 [c]	0.40 ± 0.00 [d]	0.20 ± 0.04 [e]
AH2	TSS [A]	13.64 ± 0.08 [a]	13.58 ± 0.00 [a]	13.54 ± 0.08 [a]	13.64 ± 0.08 [a]	13.64 ± 0.08 [a]
	TTA [B]	1.08 ± 0.01 [a]	1.06 ± 0.01 [b]	1.03 ± 0.01 [c]	1.08 ± 0.01 [a]	0.97 ± 0.01 [d]
	TSS/TTA	12.62	12.81	13.14	12.62	14.06
	pH	2.91 ± 0.01 [d]	2.97 ± 0.02 [c]	2.98 ± 0.00 [c]	3.19 ± 0.02 [b]	3.31 ± 0.01 [a]
	L-ascorbic acid [C]	4.23 ± 0.02 [a]	2.58 ± 0.01 [b]	1.45 ± 0.01 [c]	1.14 ± 0.01 [d]	0.96 ± 0.05 [e]
AH3	TSS [A]	13.81 ± 0.08 [a]	13.95 ± 0.08 [a]	13.95 ± 0.07 [a]	13.75 ± 0.08 [a]	13.95 ± 0.08 [a]
	TTA [B]	1.34 ± 0.01 [a]	1.33 ± 0.01 [a]	1.21 ± 0.01 [d]	1.27 ± 0.01 [b]	1.24 ± 0.01 [c]
	TSS/TTA	10.30	10.48	11.52	10.82	11.25
	pH	2.86 ± 0.01 [c]	2.90 ± 0.00 [bc]	2.94 ± 0.01 [b]	2.90 ± 0.01 [bc]	3.25 ± 0.01 [a]
	L-ascorbic acid [C]	8.07 ± 0.05 [a]	6.10 ± 0.04 [b]	3.47 ± 0.02 [c]	2.89 ± 0.57 [c]	2.14 ± 0.01 [d]

[A] TSS total soluble solids [°Brix]; [B] TTA total titratable acidity [g of malic acid/100 mL]; [C] L-ascorbic acid [mg/100 ml]; [a-e] Means with the same letter did not differ significantly.

Polysaccharides may reduce the perception of tart taste. Astringent properties of plant extracts rich in polyphenols reduced the use of polydextrose [24] and carboxymethylcellulose [25]. The reduction of the astringent effect may be the result of the adsorption of polyphenols on the surface of polysaccharides [25]. Apples are known for their polysaccharide content, especially from the pectic polysaccharides group. It has been shown that polysaccharides contained in apple tissue can interact with polyphenols through hydrophobic interactions and hydrogen bonds [26]. Mixing blue honeysuckle berry juice containing a high content of polyphenols (767.88 mg/100 mL) with apple juice with a high content of polysaccharides could lead to mutual interactions between polysaccharides and polyphenolic compounds, and thus give an effect that reduces the astringency, bitterness, and acidity of H juice. This means that contrary to the earlier quoted studies [24,25], the reduction of unfavorable astringency and bitterness can be achieved not only by adding specific polysaccharides, but also by mixing juices from raw materials differing in the content of these ingredients.

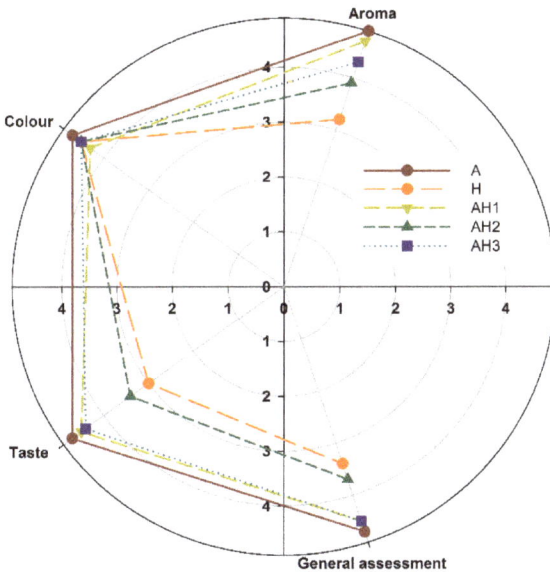

Figure 2. Sensory evaluation of mixed juices; apple and blue honeysuckle berry juice.

A number of compounds have been identified in the blue honeysuckle berry fruit that can affect the perception of the taste of this raw material. In the study of Wojdyło et al. [27], it was shown that blue honeysuckle berry fruit contains a high content of proanthocyanidins, and therefore, compounds belonging to the tannin group. Proanthocyanidins are compounds with proven effects on sensory characteristics of food. It has been shown that above all, they are compounds responsible for astringency, and they can also shape bitterness and sourness of plant raw materials [28]. In addition, recent studies have shown that iridoids (compounds from the monoterpenes group) are present in the blue honeysuckle berry fruit, which, as plant defense substances, give bitterness to plant raw materials [10,29,30]. In addition, chlorogenic acids are the dominant group of phenolic acids in blue honeysuckle berry fruits, and it has been shown that these secondary metabolites shape the astringency and bitterness of a coffee beverage [26,31].

Thus, this study disproved the notion that the higher the percentage of the more acidic juice and the lower the TSS/TTA ratio, the lesser the sensory acceptability. Therefore, further research should be aimed at understanding the effect of the percentage shares of individual juices in mixed juices on sensory characteristics and the interactions between the components of juices. This is important to improve the health benefits and sensory characteristics in the design of functional foods [32–35].

3.2. L-Ascorbic Acid Content

Vitamin C is a well-known and essential vitamin in the human diet. It acts as a natural antioxidant, preventing oxidative stress in the body [36]. Vitamin C consists of L-ascorbic acid and dehydroascorbic acid, an oxidized form of ascorbic acid [37]. In this study, the content of L-ascorbic acid in all of the juice variants was determined, and significant differences were observed (Table 2). Immediately after production, the highest content of L-ascorbic acid was observed in the H juice and the lowest in the A juice (32.59 and 0.52 mg/100 mL, respectively). In the case of the mixed juices (AH1, AH2, AH3), the higher the content of H juice, the higher the content of L-ascorbic acid found. However, after 4 months of storage, a high degree of acid degradation was observed. The lowest loss was found in AH3 juice, in which 27% of the initial content of L-ascorbic acid remained, and the highest loss was found in AH1 juice, in which only 18% of the initial content remained. The factors that caused the degradation of L-ascorbic acid observed in this study may be oxygen residue and high temperature maintained during juice pasteurization [38,39]. Vitamin C is a component with functional properties, but unfortunately, it is very sensitive to the thermal treatment applied during processing [39,40].

3.3. Anthocyanin Content

Anthocyanins are natural plant pigments that belong to the family of polyphenolic compounds classified under flavonoids. Plant raw materials containing anthocyanins are often red, pink, blue, or black in color [41,42]. Anthocyanins are also bioactive compounds that are beneficial to humans [31]. An advantage of the blue honeysuckle berry is a very high content of anthocyanins, ranging from 400 to 1500 mg/100 g [43]. Many studies have indicated that anthocyanins are health-promoting compounds, due to their antioxidant, anticancer, neuroprotective, and cardiovascular-supporting properties, among others [42,44–47]. In contrast, apples have a lesser amount of anthocyanins, and depending on the variety, the content of these compounds may vary from 0 to around 27 mg/100 g [22,23]. In our study, anthocyanins were not identified in juice A (Table 3).

This may have been due to the fact that anthocyanins are present in apple skin, which is removed to a large extent as a by-product after the pressing process, and thus, even though these compounds are present in the fruits, they might not be transferred to the juice. Therefore, the addition of blue honeysuckle berry juice may be an excellent option to enrich the apple juice with prohealth anthocyanins. The content of anthocyanins in the H juice measured immediately after production was, on average, 595.39 mg/100 mL. The higher the addition of the blue honeysuckle berry juice, the higher the average anthocyanin content detected in the mixed juices. Immediately after mixing, the highest content of anthocyanins was recorded in the AH3 juice (186.37 mg/100 mL) and the lowest in the AH1 juice (49.86 mg/100 mL) among the mixed juices. During the 4 months of storage, the total anthocyanin content was found to be reduced in the juices. The AH3 juice contained only 45% of the initial anthocyanin content, while the AH1 juice contained only 34%. However, in the H juice, 64% of the initial anthocyanin content remained. Thus, the higher the share of blue honeysuckle berry juice, the higher the total anthocyanin content and the lower the losses during storage. Anthocyanins are compounds, the stability of which is determined by the pH value [11].

A higher level of retention of total anthocyanins in the AH3 juice may be due to the fact that its pH value (below 3.0) was favorable for anthocyanins' stability during the whole storage period. In our study, six types of anthocyanins were identified: Cyanidin-3,5-O-diglucoside, cyanidin-3-O-glucoside, and cyanidin-3-O-rutinoside, peonidin-3-O-rutinoside and peonidin-3-O-glucoside, and pelargonidin-3-O-glucoside. Among these anthocyanins, the most abundant in the analyzed juices was cyanidin-3-O-glucoside. This is in accordance with other studies which showed that cyanidin-3-O-glucoside is the anthocyanin found in the highest amount in the blue honeysuckle berry [11]. Its content in the AH1 and AH3 juices was estimated to be 34.33 and 143.96 mg/100 mL, respectively. During the storage, the content of cyanidin-3-O-glucoside decreased, and after 4 months of storage, there remained only 48% and 36% of the initial content in AH3 and AH1 juices, respectively. This phenomenon could be explained by copigmentation. The effectiveness of a copigmentation reaction in reducing anthocyanin degradation depends on the type and concentration of compounds involved in the reaction. Depending on the dye and copigment structure, pH, temperature, and storage time, the course of copigmentation changes, which leads to an increase in absorbance and increases the color intensity [48].

Table 3. Content of anthocyanins [mg/100 mL] in the tested juices immediately after production and during 4 months of storage.

Juice	Parameters	Time of Storage				
		After Production	1 Month	2 Months	3 Months	4 Months
A	Total anthocyanins	0.0 ± 0.0 [a]	0.0 ± 0.0 [a]	0.0 ± 0.0 [a]	0.0 ± 0.0 [a]	0.0 ± 0.0 [a]
H	Cyanidin 3,5-O-diglucoside	60.27 ± 0.47 [a]	47.93 ± 0.51 [b]	45.75 ± 0.25 [c]	35.02 ± 1.44 [d]	30.88 ± 0.15 [e]
	Cyanidin 3-O-glucoside	460.26 ± 2.60 [a]	359.38 ± 2.98 [b]	340.18 ± 3.40 [c]	337.68 ± 12.63 [c]	305.54 ± 1.62 [d]
	Cyanidin 3-O-rutinoside	35.19 ± 0.31 [a]	29.34 ± 0.35 [b]	26.36 ± 0.36 [c]	25.43 ± 1.09 [c]	23.06 ± 0.15 [d]
	Pelargonidin 3-O-glucoside	9.32 ± 0.13 [a]	9.14 ± 0.06 [a,b]	8.80 ± 0.06 [b c]	8.40 ± 0.38 [c]	5.30 ± 0.03 [d]
	Peonidin 3-O-glucoside	23.71 ± 0.22 [a]	20.29 ± 0.22 [b]	18.70 ± 0.49 [c]	17.28 ± 0.72 [d]	15.38 ± 0.10 [e]
	Peonidin 3-O-rutinoside	6.63 ± 0.06 [a]	5.31 ± 0.33 [b]	4.29 ± 0.11 [c]	3.51 ± 0.04 [d]	3.38 ± 0.05 [d]
	Total anthocyanins	595.39 ± 3.77 [a]	471.39 ± 4.44 [b]	444.09 ± 3.74 [c]	427.32 ± 5.28 [d]	383.56 ± 2.09 [e]
AH1	Cyanidin 3,5-O-diglucoside	6.48 ± 0.10 [a]	5.07 ± 0.03 [b]	3.38 ± 0.05 [c]	2.95 ± 0.08 [d]	2.62 ± 0.02 [e]
	Cyanidin 3-O-glucoside	34.33 ± 0.84 [a]	17.12 ± 0.61 [b]	15.90 ± 0.16 [b]	14.15 ± 0.89 [c]	12.49 ± 0.17 [c]
	Cyanidin 3-O-rutinoside	4.56 ± 0.36 [a]	2.16 ± 0.01 [b]	1.15 ± 0.02 [c]	0.95 ± 0.02 [c]	0.81 ± 0.01 [c]
	Pelargonidin 3-O-glucoside	1.13 ± 0.02 [a]	1.15 ± 0.01 [a]	1.09 ± 0.01 [b]	0.96 ± 0.01 [c]	0.58 ± 0.01 [d]
	Peonidin 3-O-glucoside	3.17 ± 0.07 [a]	1.34 ± 0.04 [b]	1.29 ± 0.01 [b]	0.62 ± 0.01 [c]	0.34 ± 0.01 [d]
	Peonidin 3-O-rutinoside	0.18 ± 0.01 [a]	0.16 ± 0.00 [b]	0.15 ± 0.00 [b c]	0.15 ± 0.01 [c]	0.10 ± 0.00 [d]
	Total anthocyanins	49.86 ± 1.36 [a]	27.00 ± 0.72 [b]	22.97 ± 1.92 [c]	19.77 ± 1.03 [d]	16.94 ± 0.33 [e]
AH2	Cyanidin 3,5-O-diglucoside	9.37 ± 0.08 [b]	10.10 ± 0.10 [a]	5.95 ± 0.04 [c]	5.90 ± 0.05 [c]	5.59 ± 0.10 [d]
	Cyanidin 3-O-glucoside	68.76 ± 0.41 [a]	31.83 ± 0.16 [b]	30.63 ± 0.31 [c]	27.62 ± 0.18 [d]	25.55 ± 0.13 [e]
	Cyanidin 3-O-rutinoside	5.04 ± 0.03 [a]	4.82 ± 0.03 [b]	2.86 ± 0.02 [c]	2.65 ± 0.02 [d]	2.62 ± 0.01 [d]
	Pelargonidin 3-O-glucoside	1.20 ± 0.01 [a]	1.30 ± 0.07 [a]	0.75 ± 0.02 [c]	0.79 ± 0.15 [b]	0.80 ± 0.05 [b]
	Peonidin 3-O-glucoside	3.28 ± 0.02 [a]	2.85 ± 0.03 [b]	1.55 ± 0.01 [c]	1.41 ± 0.07 [d]	1.29 ± 0.07 [e]
	Peonidin 3-O-rutinoside	0.58 ± 0.00 [a]	0.54 ± 0.00 [b]	0.39 ± 0.00 [c]	0.36 ± 0.00 [d]	0.27 ± 0.00 [e]
	Total anthocyanins	88.22 ± 0.52 [a]	51.45 ± 0.38 [b]	42.12 ± 0.22 [c]	38.73 ± 0.34 [d]	36.11 ± 0.35 [e]
AH3	Cyanidin 3,5-O-diglucoside	20.17 ± 0.19 [a]	18.59 ± 0.11 [b]	12.31 ± 0.06 [c]	11.55 ± 0.07 [d]	6.74 ± 0.08 [e]
	Cyanidin 3-O-glucoside	143.96 ± 0.96 [a]	91.77 ± 0.58 [b]	87.87 ± 0.87 [c]	84.79 ± 0.51 [c]	68.89 ± 2.09 [d]
	Cyanidin 3-O-rutinoside	11.12 ± 0.06 [a]	10.76 ± 0.07 [b]	7.65 ± 0.05 [c]	7.38 ± 0.15 [d]	4.43 ± 0.08 [e]
	Pelargonidin 3-O-glucoside	2.71 ± 0.03 [a]	2.76 ± 0.11 [a]	2.03 ± 0.06 [b]	1.78 ± 0.10 [c]	1.15 ± 0.01 [d]
	Peonidin 3-O-glucoside	7.51 ± 0.04 [a]	6.73 ± 0.09 [a]	4.51 ± 0.09 [b]	3.94 ± 0.16 [b]	2.78 ± 0.82 [c]
	Peonidin 3-O-rutinoside	0.90 ± 0.03 [a]	0.88 ± 0.01 [a]	0.50 ± 0.06 [b]	0.43 ± 0.01 [b c]	0.39 ± 0.00 [c]
	Total anthocyanins	186.37 ± 1.3 [a]	131.49 ± 0.95 [b]	114.87 ± 1.11 [c]	109.87 ± 0.99 [d]	84.38 ± 3.08 [e]

[a–e] Means with the same letter did not differ significantly.

3.4. Total Polyphenols and Antioxidant Activity

The polyphenol content of juice depends mostly on the raw material used. In the present study, it was found that the content of TP differed significantly in the juice variants. One of the aims of this study was to enrich the apple juice, which is poorer in polyphenols (A juice—47.39 mg/100 mL) with blue honeysuckle berry juice, which is rich in polyphenols (H juice—767.88 mg/100 mL).

Enrichment with H juice increased the TP value of A juice (by 2.7, 3.5, and 6.1 times in AH1, AH2, and AH3, respectively). It was observed that the storage time also had a significant effect on the TP value of all the juices tested (Table 4).

After 4 months of storage, the content of polyphenols was 28.65 and 635.80 mg/100 mL in the A and H juices, respectively. Among the mixed juices, the AH3 juice was characterized by the highest content of TP after 4 months of storage (174.60 mg/100 mL). Additionally, the highest losses of TP content were recorded in the AH3 juice (only 61% of the initial content remained), whereas 67% and 76% of TP content remained in the AH1 and AH2 juices, respectively. Nevertheless, despite the highest losses during the 4 months of storage, the AH3 juice had 2.0 and 1.4 times higher TP content in comparison to the AH1 and AH2 juices, respectively. Enrichment by blue honeysuckle berry juice also contributed to the increase of antioxidant activity in apple juice.

Table 4. Total polyphenols (TP) [mg/100 mL] and antioxidant activity (ABTS) [µmol Trolox/mL] in juices immediately after production and during 4 months of storage.

Juice	Parameters	Time of Storage				
		After Production	1 Month	2 Months	3 Months	4 Months
A	TP	47.39 ± 0.47 [a]	42.80 ± 0.43 [b]	41.28 ± 0.41 [c]	32.21 ± 0.32 [d]	28.65 ± 0.29 [e]
	ABTS	3.94 ± 0.00 [a]	3.54 ± 0.02 [b]	3.17 ± 0.03 [c]	2.61 ± 0.18 [d]	2.26 ± 0.04 [e]
H	TP	767.88 ± 7.68 [a]	762.80 ± 7.63 [a]	750.10 ± 7.50 [a]	706.92 ± 7.07 [b]	635.80 ± 6.36 [c]
	ABTS	96.44 ± 0.48 [a]	95.24 ± 0.47 [a,b]	92.68 ± 1.11 [b]	88.15 ± 1.87 [c]	61.90 ± 0.57 [d]
AH1	TP	128.07 ± 0.76 [a]	115.98 ± 1.16 [b]	108.98 ± 1.09 [c]	94.07 ± 0.94 [d]	85.61 ± 0.86 [e]
	ABTS	20.79 ± 0.56 [a]	17.31 ± 0.23 [b]	10.87 ± 0.88 [c]	10.22 ± 0.12 [c,d]	8.95 ± 0.19 [d]
AH2	TP	165.50 ± 4.47 [a]	143.77 ± 1.44 [b]	135.41 ± 1.35 [c]	129.45 ± 1.29 [c,d]	126.00 ± 1.26 [d]
	ABTS	24.8 ± 0.76 [a]	19.25 ± 1.24 [b]	15.41 ± 0.50 [c]	13.75 ± 0.62 [c]	10.68 ± 0.08 [d]
AH3	TP	287.54 ± 2.37 [a]	258.71 ± 1.98 [b]	193.70 ± 1.93 [c]	186.07 ± 1.86 [d]	174.60 ± 1.76 [e]
	ABTS	30.09 ± 0.15 [a]	27.52 ± 0.31 [b]	26.46 ± 0.13 [c]	24.49 ± 0.30 [d]	22.09 ± 0.33 [e]

[a–e] Means with the same letter did not differ significantly.

An increase in the free radical-capturing capacity by 5.3, 6.2, and 7.6 times was observed in AH1, AH2, and AH3 juices, respectively, after production in comparison to the A juice. Immediately after production, the antioxidant activity was determined to be 20.79 and 30.09 µmol Trolox/mL in the AH1 and AH3 juices, respectively. During storage, the antioxidant activity decreased similar to the content of TP. The research of Lachowicz and Oszmiański [18] also proved that the addition of cranberrybush juice to pear juice resulted in a significant increase in its TP content and an improvement in antioxidant properties. This confirms that the use of fruit juices which are sensorially less acceptable, but having a high antioxidant capacity, even in small amounts, in the production of mixed juices based on juices from popular fruits (e.g., apples or pears) enables obtaining a product more abundant in polyphenols with potential functional properties.

3.5. Juice Color Parameters

Color is an essential quality characteristic that influences the acceptability of fruit juices [12]. It was found that the color parameters L*, a*, and b* significantly differed among the studied juices (Table 5). Immediately after production, the value of L* parameter, indicating the brightness level, was measured as 4.47 and 98.28 in the H and A juice, respectively. In the case of mixed juices (AH1, AH2, AH3), the higher the addition of H juice, the lower the value of the L* parameter and thus the lower the brightness of the tested juices. In addition, a significant effect of the storage time on the value of L* parameter was found in all the juices except the A variant. As the storage time passed, an increase in brightness was observed in the AH1, AH2, AH3, and H juices.

Immediately after production, the value of the a* parameter was measured as 0.21 and 65.79 in the A and AH3 juices, respectively. During 4 months of storage, there was a decrease in the value of this parameter in all the tested juices, with the exception of the juice A. The decrease in the share of red color resulted from the degradation of anthocyanins, which determine the color of berry juices [49]. Immediately after production, the values of the parameter b* were measured as 1.63 and 39.22 in the juices A and AH3, respectively. Moreover, during storage, the share of yellow color increased in all the studied juices. After 4 months of storage, the juices analyzed were brighter and more yellow and less red in color than the juices analyzed immediately after production. The parameter ΔE indicated the change in color during storage and expressed the possibility of distinguishing the difference in color by the human eyesight. It is considered that when ΔE value is less than 1, the color difference is not perceived; when ΔE is between 1 and 2, the color change is noticed by an experienced observer; when ΔE is between 2 and 3.5, the color change might be perceived by the consumer; when ΔE is between 3.5 and 5, the consumer can observe a clear color difference between the products; and when ΔE is above 5, the consumer has the impression of a completely distinct color [18]. After 1 month of storage, the ΔE values of the juices A and H reached 0.43 and 0.5, respectively, which meant that there was no change in color in these juice variants. On the other hand, in the case of mixed juices, the value of ΔE was 0.55 and 3.09 in the AH3 and AH2 juices, respectively. After 4 months of storage, the ΔE value of only the A juice did not exceed 2 (1.91), whereas in the other tested juices, the ΔE value exceeded 5, which indicates that after 4 months of storage, the color of the juice differed significantly from the color exhibited by the juices immediately after production. This is in line with the findings reported by other researchers that storage is one of the main determinants affecting the color of juices [50,51]. For instance, in the study of Lachowicz and Oszmiański [18], it was found that after 5 months of storage at 25 °C, an increase in ΔE was found in the pear juices mixed with cranberrybush juice.

Table 5. Colour parameters of the tested juices immediately after production and during 4 months of storage.

Juice	Parameters	Time of Storage				
		After Production	1 Month	2 Months	3 Months	4 Months
A	L^*	98.28 ± 0.98 [a]	98.28 ± 0.98 [a]	98.35 ± 0.98 [a]	98.61 ± 0.99 [a]	98.87 ± 0.99 [a]
	a^*	0.21 ± 0.00 [a]	0.17 ± 0.00 [b]	0.21 ± 0.00 [a]	0.17 ± 0.00 [b]	0.15 ± 0.00 [c]
	b^*	1.63 ± 0.02 [d]	2.06 ± 0.02 [c]	2.86 ± 0.03 [b]	3.52 ± 0.04 [a]	3.45 ± 0.03 [a]
	ΔE	-	0.43	1.23	1.92	1.91
H	L^*	4.47 ± 0.04 [d]	4.71 ± 0.05 [c]	5.44 ± 0.05 [b]	6.30 ± 0.06 [a]	6.40 ± 0.06 [a]
	a^*	35.34 ± 0.35 [a]	35.11 ± 0.35 [a]	32.97 ± 0.33 [b]	30.29 ± 0.30 [c]	29.22 ± 0.29 [d]
	b^*	7.66 ± 0.08 [d]	8.03 ± 0.08 [c]	9.35 ± 0.09 [b]	10.81 ± 0.11 [a]	10.91 ± 0.11 [a]
	ΔE	-	0.50	3.07	6.23	7.19
AH1	L^*	49.59 ± 0.50 [c]	50.76 ± 0.51 [bc]	51.81 ± 0.52 [b]	51.96 ± 0.52 [b]	53.48 ± 0.53 [a]
	a^*	57.98 ± 0.70 [a]	58.28 ± 0.58 [a]	56.95 ± 0.57 [a]	57.29 ± 0.57 [a]	55.16 ± 0.55 [b]
	b^*	10.04 ± 0.10 [e]	11.78 ± 0.12 [d]	14.48 ± 0.14 [c]	20.22 ± 0.20 [b]	33.77 ± 0.34 [a]
	ΔE	-	2.12	5.07	10.48	24.21
AH2	L^*	33.35 ± 0.33 [d]	34.83 ± 0.35 [c]	34.84 ± 0.35 [c]	36.39 ± 0.36 [b]	38.55 ± 0.39 [a]
	a^*	60.05 ± 3.44 [a]	60.38 ± 0.60 [a]	55.27 ± 0.55 [b]	52.97 ± 0.53 [bc]	49.59 ± 0.50 [c]
	b^*	29.64 ± 0.30 [e]	32.33 ± 0.32 [d]	37.92 ± 0.38 [c]	46.04 ± 0.46 [b]	59.34 ± 0.59 [a]
	ΔE	-	3.09	9.68	18.12	31.92
AH3	L^*	24.10 ± 0.24 [c]	24.29 ± 0.38 [c]	25.39 ± 0.25 [b]	26.39 ± 0.26 [a]	27.26 ± 0.27 [a]
	a^*	65.79 ± 0.66 [a]	65.28 ± 1.77 [a]	61.57 ± 0.62 [b]	62.30 ± 0.70 [b]	60.95 ± 0.71 [b]
	b^*	39.22 ± 0.39 [d]	39.31 ± 0.42 [d]	42.69 ± 0.42 [c]	44.91 ± 0.44 [b]	46.88 ± 0.46 [a]
	ΔE	-	0.55	5.61	7.06	9.60

[a–e] Means with the same letter did not differ significantly.

4. Conclusions

The results of this study indicate that the addition of blue honeysuckle berry juice can enrich the apple juice with anthocyanins, and thus improve its prohealth properties. Mixing with apple juice is also a very good way to utilize blue honeysuckle berry juice that is otherwise unpreferred by consumers due to its very high level of acidity. An advantage of the blue honeysuckle juice is undoubtedly its higher content of anthocyanins than in juices obtained from other popular fruits. This makes blue honeysuckle berry particularly valuable in the production of functional juices. To the best of our knowledge, this work is the first to deal with the influence of different doses of blue honeysuckle berry juice on the quality characteristics of mixed apple–blue honeysuckle berry juices. Therefore, this study is crucial in the current pursuit of new raw materials for the production of functional foods and in the future design of similar products on an industrial scale.

Author Contributions: A.G. and M.K. collected and reviewed the literature and wrote the manuscript (drafted and critically reviewed). A.G. and S.K. performed the experiments. A.G. and M.K. compiled the results. M.K. and S.K. reviewed the manuscript and provided supervision

Funding: This research received no external funding.

Conflicts of Interest: The authors declare no conflict of interest.

References

1. Snyder, F.; Ni, L. Chinese apples and the emerging world food trade order: Food safety, international trade, and regulatory collaboration between China and the European Union. *Chin. J. Comp. Law (CJCL)* **2017**, *5*, 253–307. [CrossRef]
2. Oliveira, B.G.; Tosato, F.; Folli, G.S.; de Leite, J.A.; Ventura, J.A.; Endringer, D.C.; Filgueiras, P.R.; Romão, W. Controlling the quality of grape juice adulterated by apple juice using ESI (-) FT-ICR mass spectrometry. *Microchem. J.* **2019**, *104033*. [CrossRef]
3. Persic, M.; Mikulic-Petkovsek, M.; Slatnar, A.; Veberic, R. Chemical composition of apple fruit, juice and pomace and the correlation between phenolic content, enzymatic activity and browning. *LWT Food Sci. Technol.* **2017**, *82*, 23–31. [CrossRef]
4. Barreira, J.C.; Arraibi, A.A.; Ferreira, I.C. Bioactive and functional compounds in apple pomace from juice and cider manufacturing: Potential use in dermal formulations. *Trends Food Sci. Technol.* **2019**, *90*, 76–87. [CrossRef]
5. Senica, M.; Stampar, F.; Veberic, R.; Mikulic-Petkovsek, M. Cyanogenic glycosides and phenolics in apple seeds and their changes during long term storage. *Sci. Hortic.* **2019**, *255*, 30–36. [CrossRef]
6. Molina, A.K.; Vega, E.N.; Pereira, C.; Dias, M.I.; Heleno, S.A.; Rodrigues, P.; Fernandes, I.F.; Barreiro, M.F.; Kostić, M.; Soković, M.; et al. Promising antioxidant and antimicrobial food colourants from *Lonicera caerulea* L. var. *Kamtschatica*. *Antioxidants* **2019**, *8*, 394. [CrossRef]
7. Auzanneau, N.; Weber, P.; Kosińska-Cagnazzo, A.; Andlauer, W. Bioactive compounds and antioxidant capacity of *Lonicera caerulea* berries: Comparison of seven cultivars over three harvesting years. *J. Food Compos. Anal.* **2018**, *66*, 81–89. [CrossRef]
8. Becker, R.; Szakiel, A. Phytochemical characteristics and potential therapeutic properties of blue honeysuckle *Lonicera caerulea* L. (*Caprifoliaceae*). *J. Herb. Med.* **2019**, *16*, 100237. [CrossRef]
9. Senica, M.; Stampar, F.; Mikulic-Petkovsek, M. Blue honeysuckle (*Lonicera cearulea* L. subs. *edulis*) berry; A rich source of some nutrients and their differences among four different cultivars. *Sci. Hortic.* **2019**, *238*, 215–221. [CrossRef]
10. Oszmiański, J.; Kucharska, A.Z. Effect of pre-treatment of blue honeysuckle berries on bioactive iridoid content. *Food Chem.* **2018**, *240*, 1087–1091. [CrossRef]
11. Rupasinghe, H.V.; Arumuggam, N.; Amararathna, M.; De Silva, A.B.K.H. The potential health benefits of haskap (*Lonicera caerulea* L.): Role of cyanidin-3-O-glucoside. *J. Funct. Foods* **2018**, *44*, 24–39. [CrossRef]
12. Wojdyło, A.; Teleszko, M.; Oszmiański, J. Physicochemical characterisation of quince fruits for industrial use: Yield, turbidity, viscosity and colour properties of juices. *Int. J. Food Sci. Technol.* **2014**, *49*, 1818–1824. [CrossRef]
13. Goiffon, J.-P.; Mouly, P.P.; Gaydou, E.M. Anthocyanic pigment determination in red fruit juices, concentrated juices and syrups using liquid chromatography. *Anal. Chim. Acta* **1999**, *382*, 39–50. [CrossRef]
14. Oszmiański, J.; Wojdyło, A. Effects of blackcurrant and apple mash blending on the phenolics contents, antioxidant capacity, and colour of juices. *Czech. J. Food Sci.* **2009**, *27*, 338–351. [CrossRef]
15. Gao, X.; Ohlander, M.; Jeppsson, N.; Bjork, I.; Trajkowski, V. Changes in antioxidant effects and their relationship to phytonutrients in fruits of sea buckthorn (*Hippophae rhamnoides* L.) during maturation. *J. Agric. Food Chem.* **2000**, *48*, 1485–1490. [CrossRef] [PubMed]
16. Re, R.; Pellegrini, N.; Proteggente, A.; Pannala, A.; Yang, M.; Rice-Evans, C. Antioxidant activity applying an improved ABTS radical cation decolorization assay. *Free Radic. Biol. Med.* **1999**, *26*, 1231–1237. [CrossRef]
17. Cendrowski, A.; Ścibisz, I.; Mitek, M.; Kieliszek, M. Influence of harvest seasons on the chemical composition and antioxidant activity in *Rosa rugosa* petals. *Agrochimica* **2018**, *62*, 157–165. [CrossRef]
18. Lachowicz, S.; Oszmiański, J. The influence of addition of cranberrybush juice to pear juice on chemical composition and antioxidant properties. *J. Food Sci. Technol.* **2018**, *55*, 3399–3407. [CrossRef] [PubMed]
19. Nath, P.; Varghese, E.; Kaur, C. Optimization of enzymatic maceration for extraction of carotenoids and total phenolics from sweet pepper using response surface methodology. *Indian J. Hortic.* **2015**, *72*, 547–552. [CrossRef]

20. Islam, M.; Ahmad, I.; Ahmed, S.; Sarker, A. Biochemical Composition and shelf life study of mixed fruit juice from orange & pineapple. *J. Environ. Sci. Nat. Resour.* **2014**, *7*, 227–232. [CrossRef]
21. Jaros, D.; Thamke, I.; Raddatz, H.; Rohm, H. Single-cultivar cloudy juice made from table apples: An attempt to identify the driving force for sensory preference. *Eur. Food Res. Technol.* **2009**, *229*, 51–61. [CrossRef]
22. Francini, A.; Sebastiani, L. Phenolic compounds in apple (*Malus* x *domestica* Borkh.): Compounds characterization and stability during postharvest and after processing. *Antioxidants* **2013**, *2*, 181–193. [CrossRef] [PubMed]
23. Lesschaeve, I.; Noble, A.C. Polyphenols: Factors influencing their sensory properties and their effects on food and beverage preferences. *Am. J. Clin. Nutr.* **2005**, *81*, 330S–335S. [CrossRef] [PubMed]
24. Sunarharum, W.B.; Williams, D.J.; Smyth, H.E. Complexity of coffee flavor: A compositional and sensory perspective. *Food Res. Int.* **2014**, *62*, 315–325. [CrossRef]
25. Troszyńska, A.; Narolewska, O.; Robredo, S.; Estrella, I.; Hernández, T.; Lamparski, G.; Amarowicz, R. The effect of polysaccharides on the astringency induced by phenolic compounds. *Food Qual. Prefer.* **2010**, *21*, 463–469. [CrossRef]
26. Fernandes, P.A.; Silva, A.M.; Evtuguin, D.V.; Nunes, F.M.; Wessel, D.F.; Cardoso, S.M.; Coimbra, M.A. The hydrophobic polysaccharides of apple pomace. *Carbohydr. Polym.* **2019**, *223*, 115132. [CrossRef]
27. Wojdyło, A.; Jáuregui, P.N.N.; Carbonell-Barrachina, A.A.; Oszmiański, J.; Golis, T. Variability of phytochemical properties and content of bioactive compounds in *Lonicera caerulea* L. var. *kamtschatica* berries. *J. Agric. Food Chem.* **2013**, *61*, 12072–12084. [CrossRef]
28. Rauf, A.; Imran, M.; Abu-Izneid, T.; Patel, S.; Pan, X.; Naz, S.; Silva, A.S.; Saeed, F.; Suleria, H.A.R. Proanthocyanidins: A comprehensive review. *Biomed. Pharmacother.* **2019**, *116*, 108999. [CrossRef]
29. Kucharska, A.; Sokół-Łętowska, A.; Oszmiański, J.; Piórecki, N.; Fecka, I. Iridoids, phenolic compounds and antioxidant activity of edible honeysuckle berries (*Lonicera caerulea* var. *kamtschatica* Sevast.). *Molecules* **2017**, *22*, 405. [CrossRef]
30. Dobler, S.; Petschenka, G.; Pankoke, H. Coping with toxic plant compounds–the insect's perspective on iridoid glycosides and cardenolides. *Phytochemistry* **2011**, *72*, 1593–1604. [CrossRef]
31. Jurikova, T.; Rop, O.; Mlcek, J.; Sochor, J.; Balla, S.; Szekeres, L.; Hegedusova, A.; Hubalek, J.; Adam, V.; Kizek, R. Phenolic profile of edible honeysuckle berries (genus *Lonicera*) and their biological effects. *Molecules* **2012**, *17*, 61–79. [CrossRef] [PubMed]
32. Iwatani, S.; Yamamoto, N. Functional food products in Japan: A review. *Food Sci. Hum. Wellness* **2019**, *8*, 96–101. [CrossRef]
33. Mark, R.; Lyu, X.; Lee, J.J.L.; Parra-Saldívar, R.; Chen, W.N. Sustainable production of natural phenolics for functional food applications. *J. Funct. Foods* **2019**, *57*, 233–254. [CrossRef]
34. Nazir, M.; Arif, S.; Sanaullah Khan, R.; Nazir, W.; Khalid, N.; Maqsood, S. Opportunities and challenges for functional and medicinal beverages: Current and future trends. *Trends Food Sci. Technol.* **2019**, *88*, 513–526. [CrossRef]
35. Kaur, S.; Das, M. Functional foods: An overview. *Food Sci. Biotechnol.* **2011**, *20*, 861–875. [CrossRef]
36. Zhao, C.N.; Li, Y.; Meng, X.; Li, S.; Liu, Q.; Tang, G.Y.; Gan, R.Y.; Li, H. Bin Insight into the roles of vitamins C and D against cancer: Myth or truth? *Cancer Lett.* **2018**, *431*, 161–170. [CrossRef]
37. Zümreoglu-Karan, B. The coordination chemistry of Vitamin C: An overview. *Coord. Chem. Rev.* **2006**, *250*, 2295–2307. [CrossRef]
38. Herbig, A.L.; Maingonnat, J.F.; Renard, C.M.G.C. Oxygen availability in model solutions and purées during heat treatment and the impact on vitamin C degradation. *LWT Food Sci. Technol.* **2017**, *85*, 493–499. [CrossRef]
39. Sapei, L.; Hwa, L. Study on the kinetics of vitamin C degradation in fresh strawberry juices. *Procedia Chem.* **2014**, *9*, 62–68. [CrossRef]
40. Mercali, G.D.; Jaeschke, D.P.; Tessaro, I.C.; Marczak, L.D.F. Study of vitamin C degradation in acerola pulp during ohmic and conventional heat treatment. *LWT Food Sci. Technol.* **2012**, *47*, 91–95. [CrossRef]
41. Gu, K.D.; Wang, C.K.; Hu, D.G.; Hao, Y.J. How do anthocyanins paint our horticultural products? *Sci. Hortic.* **2019**, *249*, 257–262. [CrossRef]
42. Sinopoli, A.; Calogero, G.; Bartolotta, A. Computational aspects of anthocyanidins and anthocyanins: A review. *Food Chem.* **2019**, *297*, 124898. [CrossRef] [PubMed]

43. Cavalcante Braga, A.R.; Murador, D.C.; Mendes De Souza Mesquita, L.; Vera De Rosso, V. Critical review Bioavailability of anthocyanins: Gaps in knowledge, challenges and future research. *J. Food Compos. Anal.* **2018**, *68*, 31–40. [CrossRef]
44. Bowen-Forbes, C.S.; Zhang, Y.; Nair, M.G. Anthocyanin content, antioxidant, anti-inflammatory and anticancer properties of blackberry and raspberry fruits. *J. Food Compos. Anal.* **2010**, *23*, 554–560. [CrossRef]
45. Teng, H.; Fang, T.; Lin, Q.; Song, H.; Liu, B.; Chen, L. Red raspberry and its anthocyanins: Bioactivity beyond antioxidant capacity. *Trends Food Sci. Technol.* **2017**, *66*, 153–165. [CrossRef]
46. Cassidy, A. Berry anthocyanin intake and cardiovascular health. *Mol. Aspects Med.* **2018**, *61*, 76–82. [CrossRef]
47. Medina dos Santos, N.; Berilli Batista, P.; Batista, Â.G.; Maróstica Júnior, M.R. Current evidence on cognitive improvement and neuroprotection promoted by anthocyanins. *Curr. Opin. Food Sci.* **2019**, *26*, 71–78. [CrossRef]
48. Kalisz, S.; Oszmiański, J.; Hładyszowski, J.; Mitek, M. Stabilization of anthocyanin and skullcap flavone complexes–Investigations with computer simulation and experimental methods. *Food Chem.* **2013**, *138*, 491–500. [CrossRef]
49. Muche, B.M.; Speers, R.A.; Rupasinghe, H.P.V. Storage temperature impacts on anthocyanins degradation, color changes and haze development in juice of "Merlot" and "Ruby" grapes (*Vitis vinifera*). *Front. Nutr.* **2018**, *5*, 100. [CrossRef]
50. Roidoung, S.; Dolan, K.D.; Siddiq, M. Estimation of kinetic parameters of anthocyanins and color degradation in vitamin C fortified cranberry juice during storage. *Food Res. Int.* **2017**, *94*, 29–35. [CrossRef]
51. Buvé, C.; Kebede, B.T.; De Batselier, C.; Carrillo, C.; Pham, H.T.T.; Hendrickx, M.; Grauwet, T.; Van Loey, A. Kinetics of colour changes in pasteurised strawberry juice during storage. *J. Food Eng.* **2018**, *216*, 42–51. [CrossRef]

© 2019 by the authors. Licensee MDPI, Basel, Switzerland. This article is an open access article distributed under the terms and conditions of the Creative Commons Attribution (CC BY) license (http://creativecommons.org/licenses/by/4.0/).

Article

Mitigation of Ammonia Emissions from Cattle Manure Slurry by Tannins and Tannin-Based Polymers

Thomas Sepperer [1,2], Gianluca Tondi [1,3,*], Alexander Petutschnigg [1,2], Timothy M. Young [4] and Konrad Steiner [5]

1. Forest Products Technology and Timber Construction Department, Salzburg University of Applied Sciences, Markt 136a, Kuchl 5431, Austria; thomas.sepperer@fh-salzburg.ac.at (T.S.); alexander.petutschnigg@fh-salzburg.ac.at (A.P.)
2. Salzburg Center for Smart Materials, Jakob-Haringerstraße 2a, Salzburg 5020, Austria
3. Land, Environment, Agriculture and Forestry Department, University of Padua, Viale dell'Università 16, 35020 Legnaro (PD), Italy
4. Center for Renewable Carbon, University of Tennessee, Knoxville, TN 37996, USA; tmyoung1@utk.edu
5. Höhere Bundeslehranstalt für Landwirtschaft Ursprung, Ursprungstraße 4, Elixhausen 5161, Austria; konrad.steiner@ursprung.at
* Correspondence: gianluca.tondi@unipd.it; Tel.: +39-049-827-2776

Received: 17 March 2020; Accepted: 7 April 2020; Published: 10 April 2020

Abstract: With the extensive use of nitrogen-based fertilizer in agriculture, ammonia emissions, especially from cattle manure, are a serious environmental threat for soil and air. The European community committed to reduce the ammonia emissions by 30% by the year 2030 compared to 2005. After a moderate initial reduction, the last report showed no further improvements in the last four years, keeping the 30% reduction a very challenging target for the next decade. In this study, the mitigation effect of different types of tannin and tannin-based adsorbent on the ammonia emission from manure was investigated. Firstly, we conducted a template study monitoring the ammonia emissions registered by addition of the tannin-based powders to a 0.1% ammonia solution and then we repeated the experiments with ready-to-spread farm-made manure slurry. The results showed that all tannin-based powders induced sensible reduction of pH and ammonia emitted. Reductions higher than 75% and 95% were registered for ammonia solution and cattle slurry, respectively, when using flavonoid-based powders. These findings are very promising considering that tannins and their derivatives will be extensively available due to the increasing interest on their exploitation for the synthesis of new-generation "green" materials.

Keywords: greenhouse gas; NH_3; tannin-furanic foam; liquid manure; natural polyphenol; agriculture; emission reduction

---

1. Introduction

Roughly 90% of the ammonia (NH_3) emissions in Europe are caused by different agricultural systems [1] from which about 41% in the animal sector are emitted by beef cattle [2]. According to Wang et al. [3], the estimated greenhouse gas (GHG) emission of beef cattle is around 50 kg NH_3 per animal and year. Once in atmosphere, ammonia evolves to nitrogen oxides (NOx), which seriously contribute to the destabilization of the nitrogen cycle with consequent increase of the temperature of the planet [4]. In fact, NH_3 global warming potential is estimated to be 265 times higher than CO_2 because it is a precursor of the greenhouse effect and also the ozone layer-depleting gas, nitrous oxide (N_2O) [5,6]. Other problems associated with ammonia include water pollution or eutrophication [7] and odor nuisance as well as soil contamination and acidification [8].

By 2030 the ammonia emission from agriculture within the EU25 should be reduced by 30%, as decided by the European Commission in 2005 [9]. Since the reductions registered through 2014, the last years have registered a moderate increase of ammonia concentration again [10].

In recent years, many research groups focused on the mitigation of ammonia emission in manure. The most common ways include:

- Manipulating the animal diet with feed additives such as electron receptors, dietary lipids, ionophores, and bioactive plant compounds to reduce the enteric fermentation. It was also considered to lower the crude protein intake, resulting in an overall reduction of NH_3 evolution [3,11].
- Covering the manure heap also temporary contributes to the reduction of ammonia emissions, and, hence their oxidation to NOx.
- Modifying the application method of the manure on the field.
- Lowering of the pH value of the manure below 5.5, causing a reduction of the ammonia emission by 62% as well as a reduction of methane by 68% [12].
- Mixing of the manure with different additives, such as urease inhibitor, which blocks the hydrolysis of urea and, therefore, reduces the ammonia emission [13,14]. Other additives, e.g., brown coal, increase the function of the H^+ ion through cation exchange. Humic acid acts by suppressing the hydrolysis of urea to ammonia [15]. Activated charcoal, pyrochar, or hydrochar have also shown a reduction of ammonia emission due to the adsorption of NH_4^+ ions and NH_3 [4,16,17]. Other extensively studied amendments include inorganic compounds, like lime and coal fly ash [18], alum [19], zeolite [20], or clay [21]. Most of these additives also involve the reduction of the pH.

Tannins are polyphenols which are produced by plants to protect against biotic and abiotic decay [22,23]. They have the strong capacity to complex proteins and this is the most important mechanism exploited to neutralize enzymes of attacking organisms [24,25]. This high complexation capacity is also the reason why tannins are used in vegetal tanning of leather, which is by far the largest application field for these polyphenols [26]. Tannins are gaining attractiveness for other industrial applications because of their sustainability and their high availability: Hydrolysable tannins (Figure 1a) are commonly used in oenology and pharmacology for their astringency [27–30] and condensed tannins (Figure 1b) are used also as flocculants for water treatment and as substituent for phenol in adhesives [31,32]. This latter family of compound is gaining interest also in polymer science because in recent years several high-performing flavonoid-based products were manufactured. Tannin-based wood preservatives, for instance, have shown an increase in mechanical performance and mitigation of ember time when ignited [33,34]. Tannin-based adhesives have allowed the production of particle boards for interior use with properties comparable with urea-formaldehyde bonded ones [35–37] and zero formaldehyde emission [38], while tannin monoliths were applied for carbon-gel production [39–41]. In this context, the most attractive products are the tannin foams because, with their strong insulation performance [42], their versatility [43,44], and their outstanding fire resistance [45], they aim to replace synthetic insulation materials such as polystyrene and polyurethane in the green building construction [46,47]. One of the major drawbacks for the tannin foams is that their end-life has not been completely clarified yet. Recently, tannin-foam powders and gels have been used as filter for emerging pollutant with good results [48–51], so the possibility to exploit the complexation capacity of tannins and their derivatives against ammonia is presented in this article. This feature will open new possibilities for the use of tannin-based polymers after their initial application in building insulation and it also will provide an easy-to-implement alternative to the already existing methods for reducing ammonia emission. Therefore, the aim of this study was to identify the capacity of tannin and tannin-based products to contain the ammonia emissions from liquid solutions, also evaluating the kinetics of the adsorption process. This study might also promote future research in which different tannin-rich agricultural by-products can be considered for the same purpose.

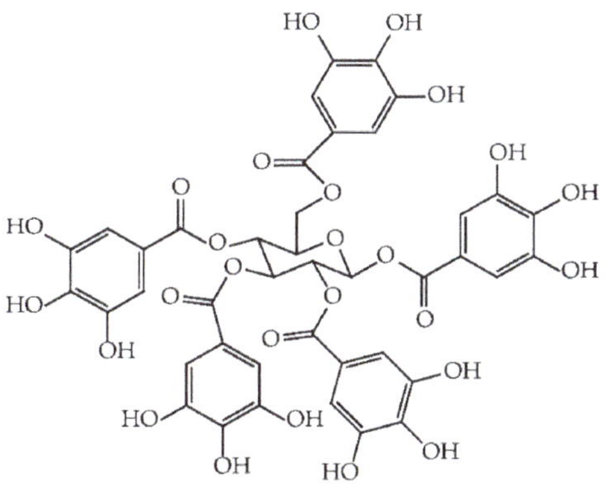

Figure 1. Example of polyphenols main molecules in: (**a**) Hydrolysable tannins, pentagalloylglucose (chestnut); (**b**) Condensed tannins, prorobinetinidin dimer (mimosa)

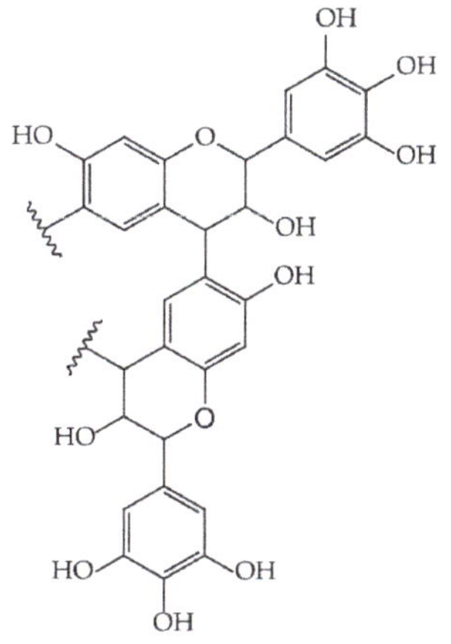

2. Materials and Methods

2.1. Chemicals

Chemicals used in the synthesis of the adsorbents were industrial mimosa tannin extract Weibull AQ (Tanac, Brazil), chestnut tannin extract (Saviolife, Italy), furfuryl alcohol (Trans Furans Chemicals, Belgium), sulphuric acid 98% (VWR, Germany), and diethyl ether (Roth, Germany). Liquid manure was collected at Scheiblhub cattle farm in southern Germany.

2.2. Adsorbent Preparation and Characterization

Industrial mimosa and chestnut extracts as representative substances for condensed and hydrolysable tannin, respectively, were used as received. The tannin gel and the tannin-furanic foam were obtained according to previous works [44,52]. In brief, for the foam, mimosa tannin (42%), furfuryl alcohol (26%), water (8%), blowing agent (6%), and sulfuric acid catalyst (18%, diluted to 32%) were mixed and cured in a hot press; for the tannin-gel, mimosa tannin was mixed with diluted sulfuric acid and reacted for 2 h at 90 °C. The tannin-based materials were dried 1 h at 103 °C and then grinded to fine powder (particle size < 125 μm) before being conditioned at 20 °C and 65% relative humidity for 1 day.

2.3. The pH Measurements

The pH of the 0.1% ammonia solution and that of the manure slurry was measured after adding the adsorbent with an Inolab pH 720 (WTW, Weilheim, Germany) pH meter. The measurement was repeated 3 times.

2.4. Adsorption Experiments

Emission of ammonia either from aqueous solution or from liquid manure was monitored in a test chamber, shown in Figure 2. The test setup consisted of a glass box with dimension of $150 \times 150 \times 70$ mm^3, a petri dish where 1 mL of ammonia-emitting solution was inserted and an ammonia sensor (type MQ-137, Vastmind LLC, Wilmington, NC, USA) sensitive until 500 parts per million (ppm), connected to an Arduino micro controller, which monitored the concentration of free NH_3 (and other volatile gasses) in the test chamber.

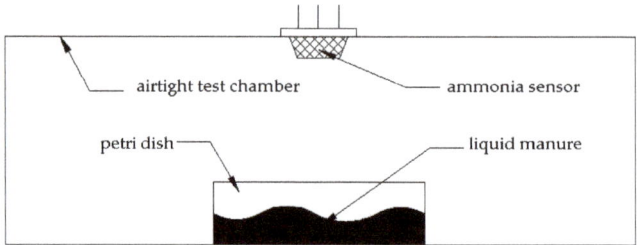

Figure 2. Setup of the test chamber to monitor emissions.

For the preliminary study on aqueous ammonia solution, 1mL of a 0.1% NH_4OH was placed in the test chamber and the emissions were monitored for one hour. Concentrations of adsorbent of 1%, 5%, and 10% by weight were added to the ammonia solution (e.g., 1 g of ammonia-emitting solution was added of 0.01, 0.05, and 0.1 g of adsorbent, respectively) and the emission was recorded for one hour.

For the adsorption experiments with liquid manure, 7.5% solid content was at first diluted 1:1 with deionized water to allow an effective mixing and homogeneous distribution of the slurry in the petri dish [4]. Also in this case, 1 mL of manure (50%) was added with 1%, 5%, and 10% by weight of adsorbent. The experiments were repeated five times.

2.5. Attenuated Total Reflectance Fourier-Transform Infrared (ATR FT-IR) Spectroscopy

ATR-FTIR analysis was done using a Perkin-Elmer Frontier FTIR spectrometer (Perkin-Elmer, Waltham, MA, USA) equipped with an ATR Miracle. Spectra were recorded between 4000 and 600 cm^{-1} with a resolution of 4 cm^{-1} and 32 scans per spectrum. FTIR spectroscopic analysis was conducted in triplicate on the four adsorbents prior and after the experiment. The spectrum of ammonia was taken in a 25% ammonia solution.

2.6. Data Processing and Analysis

Data was analyzed using OriginPro 2019b (OriginLab, Northampton, MA, USA). Fitting and modelling of curves was conducted using the nonlinear fit function of OriginPro 2019b applying the logistic fit.

3. Results and Discussion

3.1. Adsorption from Aqueous Ammonia Solution

The preliminary study on templated ammonia solution showed that all tannin-based adsorbents significantly contributed in the reduction of the NH_3 emissions (Table 1).

The pH-value and the emissions were reduced proportionally to the amounts of adsorbent applied.

Table 1. Ammonia emission from aqueous ammonia solution with and without addition of tannin-based adsorbents (standard deviation in brackets).

Adsorbent	Concentration [%]	pH	Absolute Emission [ppm]	Relative Adsorption [%]
NH$_4$OH reference		10.78 (0.05)	15.80 (0.51)	00.0 (3.23)
Chestnut tannin	1	9.35 (0.23)	11.27 (0.21)	28.7 (1.34)
	5	8.56 (0.12)	8.20 (0.38)	48.1 (2.39)
	10	7.18 (0.09)	6.23 (0.29)	60.6 (1.84)
Mimosa tannin	1	9.67 (0.15)	8.07 (0.35)	48.9 (2.19)
	5	8.69 (0.24)	5.89 (0.21)	62.7 (1.35)
	10	8.11 (0.13)	4.59 (0.09)	70.9 (0.58)
Tannin gel	1	10.10 (0.11)	8.36 (0.18)	47.1 (1.15)
	5	8.37 (0.09)	6.70 (0.36)	57.6 (2.26)
	10	7.45 (0.25)	3.52 (0.20)	77.7 (1.28)
Tannin-foam powder	1	9.85 (0.37)	12.67 (0.24)	19.8 (1.54)
	5	8.42 (0.28)	7.38 (0.19)	53.3 (1.20)
	10	7.97 (0.16)	5.99 (0.16)	62.1 (1.00)

Despite the addition of chestnut extract decreasing the pH more consistently, the higher ammonia adsorption was observed with tannin gel. In Figure 3 the relation between pH and ammonia adsorption is reported.

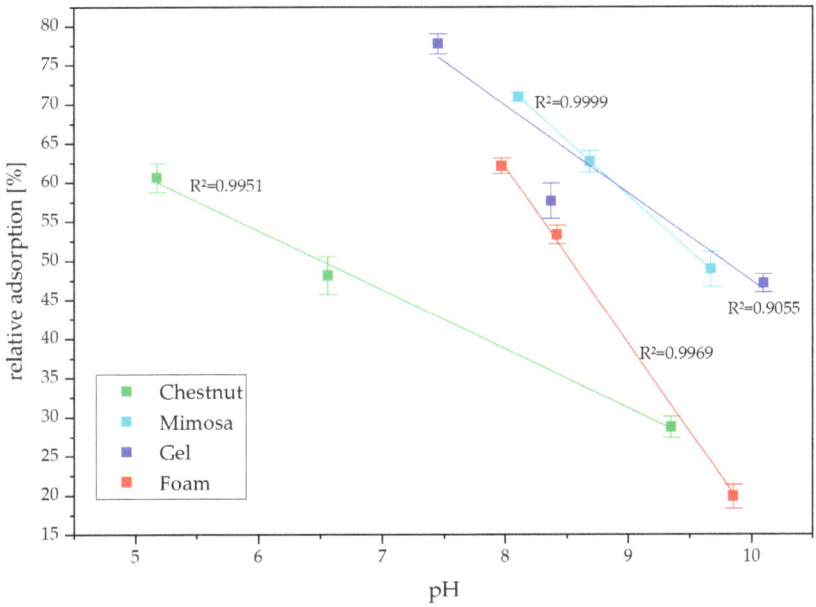

Figure 3. Influence of pH and tannin-based adsorbents on ammonia adsorption capacity.

We observed that there was a linear proportionality between pH and relative adsorption but the chestnut tannin extract depicted a gentler slope, meaning that its ammonia fixation was not as effective as the other three adsorbents. In particular, the tannin foam powder seemed to be the more effective at lower pH, even if the tannin gel reached higher absolute adsorption levels because of its higher acidity. The mimosa powder showed a similar trend suggesting that the type of flavonoid was the most important factor for the fixation of ammonia.

The FTIR spectrum of mimosa tannin can be interpreted according to [53]. In this study, the powder was measured before and after adsorption (Figure 4) and highlighted that the adsorption of ammonia significantly modified the profile of the spectrum presenting three new absorbances: (1) In the region in red between 1700 and 1630 cm^{-1} the H-N-H scissoring vibration of ammonia occurred, while (2) in the area between 1200 and 800 the N-H wagging took place [54], and (3) the clear band signal at around 1350 cm^{-1} might be due to the complexation of amino groups with tannin [55], with similar mechanism occurring for vegetable leather tanning [56].

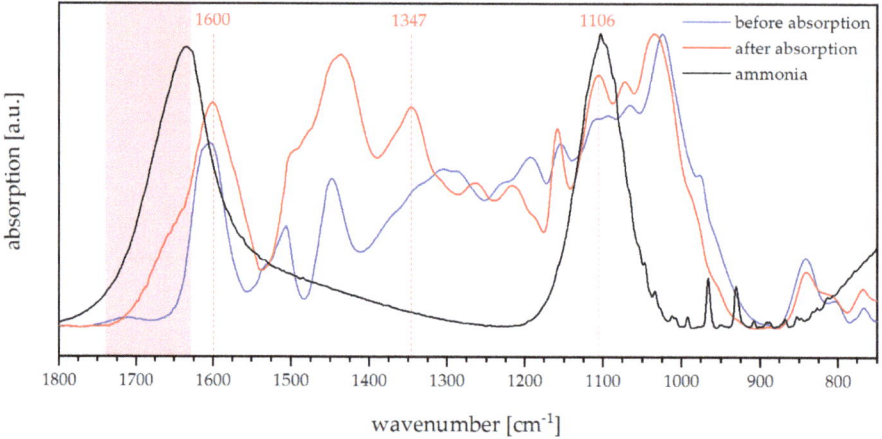

Figure 4. FTIR spectra of mimosa tannin before and after ammonia adsorption experiment.

These evidences would suggest that fixation of ammonia occurs with a process similar to the transformation of hide to leather during vegetable tanning. Although this phenomenon does not necessarily involve new covalent bonding, the complex tannin-ammonia is highly stable. Such complexation guarantees that nitrogen remains long enough in the soil, where it contributes to the fertilization also after the oxidation of ammonia to nitrites and nitrates [57].

3.2. Results of the Adsorption Experiment for Liquid Manure

The fixation of ammonia by tannin-based powders occurred successfully, but the applicability of the system had to be tested in real case. Therefore, the adsorbents were applied on 1:1 diluted manure slurry suspension, and the adsorption are registered in Table 2.

In the case of manure, the emitted ammonia was 30 times higher than in the preliminary test with the template but the efficacy of the tannin-based adsorbents was also outstanding.

In these conditions, the buffering effect of the manure did not allow consistent pH reductions but the extreme efficacy of tannin as adsorbent was confirmed. These findings also corroborated that flavonoids offer higher ammonia adsorptions than the hydrolyzed tannin. Mimosa-based adsorbent showed superior adsorption, always over 97.5%.

Interestingly, the relative ammonia adsorptions by percentage registered in the cattle manure slurry were significantly higher than the ones obtained by the adsorption of ammonia alone (Table 1). This observation suggests that the efficacy of the adsorption was improved when higher concentrations of ammonia were applied and that the presence of other substances in the manure did not affect the mitigation effect of tannins. On the other hand, we can also highlight that when the ammonia concentrations were low, tannin-based adsorbents could not completely fix them (at least, using the tested concentration of max 10%) and few ppm of ammonia were emitted anyways.

Table 2. Ammonia emission from cattle manure slurry with and without addition of tannin-based adsorbents (standard deviation in brackets).

Adsorbent	Concentration [%]	pH	Absolute Emission [ppm]	Relative Adsorption [%]
Manure reference		7.66 (0.32)	430.30 (20.1)	00.0 (4.67)
Chestnut tannin	1	7.12 (0.20)	180.30 (3.38)	58.1 (0.79)
	5	6.92 (0.13)	54.70 (7.89)	87.3 (1.83)
	10	6.61 (0.11)	44.80 (5.87)	89.6 (1.36)
Mimosa tannin	1	7.54 (0.20)	46.90 (6.05)	89.1 (1.41)
	5	7.30 (0.21)	11.50 (2.63)	97.3 (0.61)
	10	6.92 (0.15)	7.30 (1.96)	98.3 (0.46)
Tannin gel	1	7.39 (0.20)	50.10 (4.29)	88.3 (0.99)
	5	6.72 (0.21)	47.90 (4.37)	88.9 (1.01)
	10	6.27 (0.12)	4.40 (0.6)	99.0 (0.15)
Tannin-foam powder	1	7.43 (0.25)	191.00 (9.52)	55.5 (2.21)
	5	6.86 (0.11)	15.70 (3.21)	96.3 (0.75)
	10	6.17 (0.12)	10.80 (2.85)	97.5 (0.66)

The sensor was built to be selective for ammonia. However, ethanol, CO, and CH_4 may interfere with the measurement to some extent. In case this interference occurs, the ammonia value registered by the manure reference could have been too high, because part of the emission was due to the other gases. However, if this interference affected the measurements proportionally, the adsorption results in percentage would be the same. If not, the outstanding adsorption performance of tannin-based powder will be not specific, but still highly effective.

3.3. Emission Kinetics of Ammonia from Liquid Manure

Monitoring the emission during one hour, we observed that initially the slope of emission was very steep and, after around 30 min, the slope become milder until reaching a plateau after one hour (Figure 5).

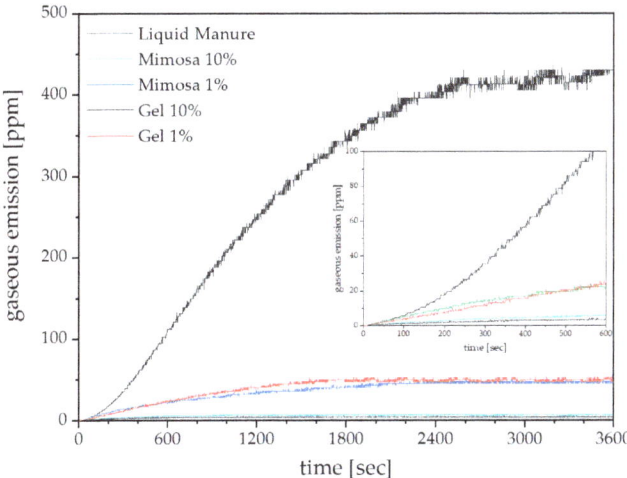

Figure 5. Evolution of NH_3 in liquid manure slurry with and without tannin-based adsorbents.

As soon as any adsorbent was added, the emissions of ammonia sank and the plateau of emission was reached sooner (around 20 min for 1% and roughly 15 minutes more for 10% concentration). These fast adsorptions allowed us to add the adsorbent just before field spreading of the slurry because relatively short storage time was needed. The tannin-based adsorbent could be added directly in the tractor tank when the slurry was loaded from the digestor. This process will guarantee a homogeneous distribution of the adsorbent and the higher viscosity will maintain the insoluble tannin fraction in suspension without need for further stirring during the spreading step.

Modelling of the logistic function (Equation (1)) to fit the experimental data for mimosa tannin and gel resulted in the parameters presented in Table 3, where A_1 and A_2 are the calculated starting and finishing values to best fit the experimental data, x_0 describes the inflection point, and p the power of the function. For liquid manure, the simulated function matched the experimental data with an accuracy of more than 99.99%. For all other simulations, correlations above 90% were observed.

$$f(x) = A_2 + \frac{(A_1 - A_2)}{1 + \left(\frac{x}{x_0}\right)^p} \tag{1}$$

Table 3. Parameters for the logistic modelling of gaseous emissions from liquid manure.

Parameter	Liquid Manure	Mimosa 10%	Mimosa 1%	Tannin Gel 10%	Tannin Gel 1%
A_1	8.243	0.857	2.377	0.612	4.552
A_2	468.000	7.457	56.514	4.593	52.093
x_0	1120.041	265.630	867.117	269.066	698.098
p	2.156	1.723	1.285	1.716	2.312
R^2	0.99996	0.94723	0.98583	0.91245	0.98183

Using the modelled equation, the rate of gaseous emission can be calculated. It was found to be quite slow for the manure mixed with adsorbents (between 0.002 and 0.05 ppm/sec at 600 seconds for 10% and 1% gel, respectively). While for the untreated manure the rate of emission ranged from 0.12 ppm/sec (at 200 s) to 0.27 ppm/sec at 600 s.

Assuming that there was no more significant ammonia emission when the experimental maximum value (allowing a 95% confidence band) was reached, one can use this equation to calculate the exact time. In the case of the manure, the calculated value was at around 1 h. When looking at the 10% gel or mimosa tannin addition, the theoretical value was around 20 minutes and for the 1% adsorbents, after 45 minutes. These values were in line with the experimentally observed ones.

4. Conclusions

In this research the mitigation effect of tannin-based adsorbents onto ammonia and gaseous emission was investigated. Chestnut and mimosa industrial tannin extracts, tannin gel, and tannin-furanic foams were tested against a 0.1% template ammonia solution and 50% diluted cattle manure slurry. Significant to outstanding reductions of ammonia emissions were registered in every test. Even small concentrations of tannin-based powders already offer good mitigation of ammonia adsorption. When 10% of adsorbent was used, the emissions of ammonia sank up to 77% for ammonia and 99% for manure. It was confirmed that the pH had an important role on the ammonia adsorption, but the type of tannin also had a major impact. Mimosa tannin-based adsorbents showed roughly 10% higher adsorptions than chestnut powder in both experiments.

The presence of adsorbent not only decreased the emission of ammonia but also shortened the time until the maximum concentration of ammonia was emitted up to around 10 times (from 30 to 3 minutes).

The FTIR measurements highlighted that ammonia was fixed in the tannin-based adsorbent: The vibrations of free NH_3 were found in the adsorbent, together with the signal at around 1350 cm^{-1}, typical for complexed amino groups in tanned leather, with this latter suggesting a stable complexation of ammonia.

These results promote tannins and tannin-based products as suitable adsorbents to fix ammonia and, hence, to fulfil the requirements of the European Union concerning the reduction of agricultural ammonia emissions. Further studies are necessary to investigate the long-term effects on the ammonia emission reduction and to clarify the transformation process of nitrogen occurring in the soil when complexed to tannin-based adsorbents.

Author Contributions: Conceptualization, T.S. and K.S.; methodology, T.S. and T.M.Y.; investigation, T.S.; resources, A.P.; writing—original draft preparation, T.S. and G.T.; writing—review and editing, G.T.; visualization, T.S.; supervision, A.P.; funding acquisition, A.P. All authors have read and agreed to the published version of the manuscript.

Funding: This research was funded by AWS (Austrian Wirtschaftsservice), grant number P1727558-IWB01 in scope of the EFRE (Europäischer Fonds zur regionalen Entwicklung) Project IWB (Investitionen in Wachstum und Beschäftigung) Zentrum Smart Materials.

Conflicts of Interest: The authors declare no conflict of interest

References

1. Erisman, J.W.; Bleeker, A.; Hensen, A.; Vermeulen, A. Agricultural air quality in Europe and the future perspectives. *Atmos. Environ.* **2008**, *42*, 3209–3217. [CrossRef]
2. Paulot, F.; Jacob, D.J.; Pinder, R.W.; Bash, J.O.; Travis, K.; Henze, D.K. Ammonia emissions in the United States, European Union, and China derived by high-resolution inversion of ammonium wet deposition data: Interpretation with a new agricultural emissions inventory (MASAGE_NH3). *J. Geophys. Res. Atmos.* **2014**, *119*, 4343–4364. [CrossRef]
3. Wang, Y.; Li, X.; Yang, J.; Tian, Z.; Sun, Q.; Xue, W.; Dong, H. Mitigating Greenhouse Gas and Ammonia Emissions from Beef Cattle Feedlot Production: A System Meta-Analysis. *Environ. Sci. Technol.* **2018**, *52*, 11232–11242. [CrossRef] [PubMed]
4. Gronwald, M.; Helfrich, M.; Don, A.; Fuß, R.; Well, R.; Flessa, H. Application of hydrochar and pyrochar to manure is not effective for mitigation of ammonia emissions from cattle slurry and poultry manure. *Biol. Fertil. Soils* **2018**, *54*, 451–465. [CrossRef]
5. Ravishankara, A.R.; Daniel, J.S.; Portmann, R.W. Nitrous Oxide (N_2O): The Dominant Ozone-Depleting Substance Emitted in the 21st Century. *Science* **2009**, *326*, 123–125. [CrossRef]
6. Zhu, M.; Lai, J.-K.; Wachs, I.E. Formation of N_2O greenhouse gas during SCR of NO with NH_3 by supported vanadium oxide catalysts. *Appl. Catal. B Environ.* **2018**, *224*, 836–840. [CrossRef]
7. Bergstrom, A.-K.; Jansson, M. Atmospheric nitrogen deposition has caused nitrogen enrichment and eutrophication of lakes in the northern hemisphere. *Glob. Chang. Biol.* **2006**, *12*, 635–643. [CrossRef]
8. Song, H.; Che, Z.; Cao, W.; Huang, T.; Wang, J.; Dong, Z. Changing roles of ammonia-oxidizing bacteria and archaea in a continuously acidifying soil caused by over-fertilization with nitrogen. *Environ. Sci. Pollut. Res.* **2016**, *23*, 11964–11974. [CrossRef]
9. Commission of the European Communities. Communication from the Commission to the Council and the European Parliament. Available online: https://ec.europa.eu/knowledge4policy/publication/commission-european-communities-communication-commission-council-european-parliament_en (accessed on 5 March 2020).
10. European Environment Agency. Ammonia Emissions from Agriculture Continue to Pose Problems for Europe. Available online: https://www.eea.europa.eu/highlights/ammonia-emissions-from-agriculture-continue (accessed on 3 March 2020).
11. Wang, Y.; Xue, W.; Zhu, Z.; Yang, J.; Li, X.; Tian, Z.; Dong, H.; Zou, G. Mitigating ammonia emissions from typical broiler and layer manure management—A system analysis. *Waste Manag.* **2019**, *93*, 23–33. [CrossRef]
12. Sommer, S.G.; Clough, T.J.; Balaine, N.; Hafner, S.D.; Cameron, K.C. Transformation of Organic Matter and the Emissions of Methane and Ammonia during Storage of Liquid Manure as Affected by Acidification. *J. Environ. Qual.* **2017**, *46*, 514–521. [CrossRef]
13. Varel, V.H.; Nienaber, J.A.; Freetly, H.C. Conservation of nitrogen in cattle feedlot waste with urease inhibitors. *J. Anim. Sci.* **1999**, *77*, 1162–1168. [CrossRef] [PubMed]

14. Hagenkamp-Korth, F.; Haeussermann, A.; Hartung, E.; Reinhardt-Hanisch, A. Reduction of ammonia emissions from dairy manure using novel urease inhibitor formulations under laboratory conditions. *Biosyst. Eng.* **2015**, *130*, 43–51. [CrossRef]
15. Chen, D.; Sun, J.; Bai, M.; Dassanayake, K.B.; Denmead, O.T.; Hill, J. A new cost-effective method to mitigate ammonia loss from intensive cattle feedlots: Application of lignite. *Sci. Rep.* **2015**, *5*, 16689. [CrossRef] [PubMed]
16. Steiner, C.; Das, K.C.; Melear, N.; Lakly, D. Reducing Nitrogen Loss during Poultry Litter Composting Using Biochar. *J. Environ. Qual.* **2010**, *39*, 1236–1242. [CrossRef] [PubMed]
17. Spokas, K.A.; Novak, J.M.; Venterea, R.T. Biochar's role as an alternative N-fertilizer: Ammonia capture. *Plant Soil* **2012**, *350*, 35–42. [CrossRef]
18. Wong, J.W.-C.; Fung, S.O.; Selvam, A. Coal fly ash and lime addition enhances the rate and efficiency of decomposition of food waste during composting. *Bioresour. Technol.* **2009**, *100*, 3324–3331. [CrossRef]
19. Shi, Y.; Parker, D.B.; Cole, N.A.; Auvermann, B.W.; Mehlhorn, J.E. Surface amendments to minimize ammonia emissions from beef cattle feedlots. *Trans. ASAE* **2001**, *44*, 677.
20. Chan, M.T.; Selvam, A.; Wong, J.W.C. Reducing nitrogen loss and salinity during 'struvite' food waste composting by zeolite amendment. *Bioresour. Technol.* **2016**, *200*, 838–844. [CrossRef]
21. Chen, H.; Awasthi, M.K.; Liu, T.; Zhao, J.; Ren, X.; Wang, M.; Duan, Y.; Awasthi, S.K.; Zhang, Z. Influence of clay as additive on greenhouse gases emission and maturity evaluation during chicken manure composting. *Bioresour. Technol.* **2018**, *266*, 82–88. [CrossRef]
22. Comandini, P.; Lerma-García, M.J.; Simó-Alfonso, E.F.; Toschi, T.G. Tannin analysis of chestnut bark samples (Castanea sativa Mill.) by HPLC-DAD–MS. *Food Chem.* **2014**, *157*, 290–295. [CrossRef]
23. Hagerman, A.E. *The Tannin Handbook*; Miami University: Oxford, OH, USA, 2002.
24. Halvorson, J.J.; Gonzalez, J.M.; Hagerman, A.E. Retention of tannin-C is associated with decreased soluble nitrogen and increased cation exchange capacity in a broad range of soils. *Soil Sci. Soc. Am. J.* **2013**, *77*, 1199–1213. [CrossRef]
25. Fukushima, M.; Yamamoto, M.; Komai, T.; Yamamoto, K. Studies of structural alterations of humic acids from conifer bark residue during composting by pyrolysis-gas chromatography/mass spectrometry using tetramethylammonium hydroxide (TMAH-py-GC/MS). *J. Anal. Appl. Pyrolysis* **2009**, *86*, 200–206. [CrossRef]
26. Grand View Research. *Tannin Market Analysis by Sources (Plants, Brown Algae), by Product (Hydrolysable, Non-Hydrolysable, Phlorotannins), by Application (Leather Tanning, Wine Production, Wood Adhesives), & Segment Forecasts, 2014–2025*; Grand View Research Inc.: San Francisco, CA, USA, 2017.
27. Bueno, F.G.; Panizzon, G.P.; de Leite Mello, E.V.S.; Lechtenberg, M.; Petereit, F.; de Mello, J.C.P.; Hensel, A. Hydrolyzable tannins from hydroalcoholic extract from Poincianella pluviosa stem bark and its wound-healing properties: Phytochemical investigations and influence on in vitro cell physiology of human keratinocytes and dermal fibroblasts. *Fitoterapia* **2014**, *99*, 252–260. [CrossRef] [PubMed]
28. Pizzi, A. Tannins: Major Sources, Properties and Applications. In *Monomers, Polymers and Composites from Renewable Resources*; Elsevier: Amsterdam, The Netherlands, 2008; pp. 179–199.
29. Buzzini, P.; Arapitsas, P.; Goretti, M.; Branda, E.; Turchetti, B.; Pinelli, P.; Ieri, F.; Romani, A. Antimicrobial and Antiviral Activity of Hydrolysable Tannins. *Mini-Reviews Med. Chem.* **2008**, *8*, 1179–1187. [CrossRef]
30. Shahat, A.A.; Marzouk, M.S. Tannins and Related Compounds from Medicinal Plants of Africa. In *Medicinal Plant Research in Africa*; Elsevier: Amsterdam, The Netherlands, 2013; pp. 479–555.
31. Lei, H.; Pizzi, A.; Du, G. Environmentally friendly mixed tannin/lignin wood resins. *J. Appl. Polym. Sci.* **2008**, *107*, 203–209. [CrossRef]
32. Moubarik, A.; Pizzi, A.; Allal, A.; Charrier, F.; Charrier, B. Cornstarch and tannin in phenol–formaldehyde resins for plywood production. *Ind. Crops Prod.* **2009**, *30*, 188–193. [CrossRef]
33. Tondi, G.; Wieland, S.; Wimmer, T.; Thevenon, M.F.; Pizzi, A.; Petutschnigg, A. Tannin-boron preservatives for wood buildings: Mechanical and fire properties. *Eur. J. Wood Wood Prod.* **2012**, *70*, 689–696. [CrossRef]
34. Sommerauer, L.; Thevenon, M.-F.; Petutschnigg, A.; Tondi, G. Effect of hardening parameters of wood preservatives based on tannin copolymers. *Holzforschung* **2019**, *73*, 457–467. [CrossRef]
35. Pizzi, A. Recent developments in eco-efficient bio-based adhesives for wood bonding: Opportunities and issues. *J. Adhes. Sci. Technol.* **2006**, *20*, 829–846. [CrossRef]
36. Pizzi, A. Tannin based Wood Adhesives Technology. In *Advanced Wood Adhesives Technology*; Decker: New York, NY, USA, 1994; p. 304. ISBN 9780824792664.

37. Pizzi, A.; Meikleham, N.; Dombo, B.; Roll, W. Autocondensation-based, zero-emission, tannin adhesives for particleboard. *Holz Roh Werkst.* **1995**, *53*, 201–204. [CrossRef]
38. Moubarik, A.; Mansouri, H.R.; Pizzi, A.; Charrier, F.; Allal, A.; Charrier, B. Corn flour-mimosa tannin-based adhesives without formaldehyde for interior particleboard production. *Wood Sci. Technol.* **2013**, *47*, 675–683. [CrossRef]
39. Szczurek, A.; Amaral-Labat, G.; Fierro, V.; Pizzi, A.; Celzard, A. The use of tannin to prepare carbon gels. Part II. Carbon cryogels. *Carbon* **2011**, *49*, 2785–2794. [CrossRef]
40. Amaral-Labat, G.; Szczurek, A.; Fierro, V.; Celzard, A. Unique bimodal carbon xerogels from soft templating of tannin. *Mater. Chem. Phys.* **2015**, *149*, 193–201. [CrossRef]
41. Rey-Raap, N.; Szczurek, A.; Fierro, V.; Celzard, A.; Menéndez, J.A.; Arenillas, A. Advances in tailoring the porosity of tannin-based carbon xerogels. *Ind. Crops Prod.* **2016**, *82*, 100–106. [CrossRef]
42. Shirmohammadli, Y.; Efhamisisi, D.; Pizzi, A. Tannins as a sustainable raw material for green chemistry: A review. *Ind. Crops Prod.* **2018**, *126*, 316–332. [CrossRef]
43. Kolbitsch, C.; Link, M.; Petutschnigg, A.; Wieland, S.; Tondi, G. Microwave Produced Tannin-furanic Foams. *J. Mater. Sci. Res.* **2012**, *1*, 84. [CrossRef]
44. Link, M.; Kolbitsch, C.; Tondi, G.; Ebner, M.; Wieland, S.; Petutschnigg, A. Formaldehyde-free tannin-based foams and their use as lightweight panels. *BioResources* **2011**, *6*, 4218–4228.
45. Tondi, G.; Pizzi, A. Tannin-based rigid foams: Characterization and modification. *Ind. Crops Prod.* **2009**, *29*, 356–363. [CrossRef]
46. Tondi, G.; Link, M.; Kolbitsch, C.; Lesacher, R.; Petutschnigg, A. Pilot plant up-scaling of tannin foams. *Ind. Crops Prod.* **2016**, *79*, 211–218. [CrossRef]
47. Eckardt, J.; Neubauer, J.; Sepperer, T.; Donato, S.; Zanetti, M.; Cefarin, N.; Vaccari, L.; Lippert, M.; Wind, M.; Schnabel, T.; et al. Synthesis and Characterization of High-Performing Sulfur-Free Tannin Foams. *Polymers* **2020**, *12*, 564. [CrossRef]
48. Sánchez-Martín, J.; Beltrán-Heredia, J.; Carmona-Murillo, C. Adsorbents from Schinopsis balansae: Optimisation of significant variables. *Ind. Crops Prod.* **2011**, *33*, 409–417. [CrossRef]
49. Sepperer, T.; Neubauer, J.; Eckardt, J.; Schnabel, T.; Petutschnigg, A.; Tondi, G. Pollutant Absorption as a Possible End-Of-Life Solution for Polyphenolic Polymers. *Polymers* **2019**, *11*, 911. [CrossRef] [PubMed]
50. Gurung, M.; Adhikari, B.B.; Kawakita, H.; Ohto, K.; Inoue, K.; Alam, S. Recovery of Au(III) by using low cost adsorbent prepared from persimmon tannin extract. *Chem. Eng. J.* **2011**, *174*, 556–563. [CrossRef]
51. Tondi, G.; Oo, C.W.; Pizzi, A.; Trosa, A.; Thevenon, M.F. Metal adsorption of tannin based rigid foams. *Ind. Crops Prod.* **2009**, *29*, 336–340. [CrossRef]
52. Sánchez-Martín, J.; González-Velasco, M.; Beltrán-Heredia, J.; Gragera-Carvajal, J.; Salguero-Fernández, J. Novel tannin-based adsorbent in removing cationic dye (Methylene Blue) from aqueous solution. Kinetics and equilibrium studies. *J. Hazard. Mater.* **2010**, *174*, 9–16. [CrossRef]
53. Tondi, G.; Petutschnigg, A. Middle infrared (ATR FT-MIR) characterization of industrial tannin extracts. *Ind. Crops Prod.* **2015**, *65*, 422–428. [CrossRef]
54. Chemical Education Digital Library Models 360. Available online: http://www.chemeddl.org/resources/models360/models.php?pubchem=6329# (accessed on 5 March 2020).
55. Akter, N.; Hossain, M.A.; Hassan, M.J.; Amin, M.K.K.; Elias, M.; Rahman, M.M.; Asiri, A.M.; Siddiquey, I.A.; Hasnat, M.A. Amine modified tannin gel for adsorptive removal of Brilliant Green dye. *J. Environ. Chem. Eng.* **2016**, *4*, 1231–1241. [CrossRef]
56. Aslan, A. Improving the Dyeing Properties of Vegetable Tanned Leathers Using Chitosan Formate. *Ekoloji* **2013**, *22*, 26–35. [CrossRef]
57. Koren, D.W.; Gould, W.D.; Bédard, P. Biological removal of ammonia and nitrate from simulated mine and mill effluents. *Hydrometallurgy* **2000**, *56*, 127–144. [CrossRef]

© 2020 by the authors. Licensee MDPI, Basel, Switzerland. This article is an open access article distributed under the terms and conditions of the Creative Commons Attribution (CC BY) license (http://creativecommons.org/licenses/by/4.0/).

Review

Tannin Gels and Their Carbon Derivatives: A Review

Flavia Lega Braghiroli [1,*], Gisele Amaral-Labat [2], Alan Fernando Ney Boss [2], Clément Lacoste [3] and Antonio Pizzi [4]

[1] Centre Technologique des Résidus Industriels (CTRI, Technology Center for Industrial Waste), Cégep de l'Abitibi-Témiscamingue (College of Abitibi-Témiscamingue), 425 Boul. du Collège, Rouyn-Noranda, QC J9X 5E5, Canada

[2] Department of Metallurgical and Materials Engineering PMT-USP, University of São Paulo, Avenida Mello Moraes, 2463, Cidade Universitária, São Paulo CEP 05508-030, Brazil; gisele.amaral@usp.br (G.A.-L.); alan.boss@usp.br (A.F.N.B.)

[3] Centre des Matériaux des Mines d'Alès (C2MA), IMT Mines d'Alès, Université de Montpellier, 6 Avenue de Clavières, 30319 Alès CEDEX, France; clement.lacoste@mines-ales.fr

[4] LERMAB-ENSTIB, University of Lorraine, 27 rue du Merle Blanc, BP 1041, 88051 Epinal, France; antonio.pizzi@univ-lorraine.fr

* Correspondence: flavia.braghiroli@cegepat.qc.ca; Tel.: +1-(819)-762-0931 (ext. 1748)

Received: 21 August 2019; Accepted: 5 October 2019; Published: 8 October 2019

Abstract: Tannins are one of the most natural, non-toxic, and highly reactive aromatic biomolecules classified as polyphenols. The reactive phenolic compounds present in their chemical structure can be an alternative precursor for the preparation of several polymeric materials for applications in distinct industries: adhesives and coatings, leather tanning, wood protection, wine manufacture, animal feed industries, and recently also in the production of new porous materials (i.e., foams and gels). Among these new polymeric materials synthesized with tannins, organic and carbon gels have shown remarkable textural and physicochemical properties. Thus, this review presents and discusses the available studies on organic and carbon gels produced from tannin feedstock and how their properties are related to the different operating conditions, hence causing their cross-linking reaction mechanisms. Moreover, the steps during tannin gels preparation, such as the gelation and curing processes (under normal or hydrothermal conditions), solvent extraction, and gel drying approaches (i.e., supercritical, subcritical, and freeze-drying) as well as the methods available for their carbonization (i.e., pyrolysis and activation) are presented and discussed. Findings from organic and carbon tannin gels features demonstrate that their physicochemical and textural properties can vary greatly depending on the synthesis parameters, drying conditions, and carbonization methods. Research is still ongoing on the improvement of tannin gels synthesis and properties, but the review evaluates the application of these highly porous materials in multidisciplinary areas of science and engineering, including thermal insulation, contaminant sorption in drinking water and wastewater, and electrochemistry. Finally, the substitution of phenolic materials (i.e., phenol and resorcinol) by tannin in the production of gels could be beneficial to both the bioeconomy and the environment due to its low-cost, bio-based, non-toxic, and non-carcinogenic characteristics.

Keywords: tannin; polyphenolic molecules; sol-gel; organic gel; carbon gel; hydrothermal carbonization; porous materials; pore structure; biopolymer; low-cost

1. The Chemistry of Tannins

1.1. Definition and Classification

The utilization of tannins by human beings dates back to the second millennium before Christ, with leather tanning. The term 'tannins' itself is etymologically derived from the ancient Keltic lexical root 'tan', meaning 'oak' in reference to leather processing. After centuries of utilization, the chemical structures of tannins, their role, and their chemistry were extensively described for nearly half of century. Tannins are the most abundant compounds from the biomass after cellulose, hemicellulose, and lignin [1]. Their chemical structure is comprised of complex and heterogeneous polyphenolic secondary metabolites, biosynthesized by higher plants, with molar weights ranging from 300 g/mol for simple phenolic compounds to over 3000 g/mol for highly polymerized structures. Tannins are present in each cytoplasm of vegetable cells, and therefore in almost every part of plants such as barks, woods, leaves, fruits, roots, and seeds [2].

However, their quantity and composition may vary according to their vegetal source such as their botanic species, age, plant tissue, and environment. According to the species, a higher concentration has been reported in the wood of quebracho (*Schinopsis balansae* and *lorentzii*) and chesnut (*Castanea* sp.), in the bark of oak (*Quercus* sp.), pine (*Pinus* sp.) and mimosa (*Acacia mearnsii* formerly *mollissima* de Wildt), or in other tissues like in fruit pods of tara (*Caesalpinia spinosa*). Process extraction parameters (solvent, additives, temperature, time) are also a key factor in the composition of tannins extracts. Thus, the heterogeneous nature of tannins makes it impossible to settle on a universal method for their extraction, and imposes a reliance on relies on its final use for obtaining extracts [3]. According to their monomer unit, two wide classes of tannins exist: (i) hydrolysable tannins such as gallotannins (gallic acid compounds and glucose) and ellagitannins (composed of biaryl units and glucose), and (ii) condensed polyflavonoid tannins, this latter being stable and rarely subject to hydrolysis [4]. Some species produce exclusively either gallotannins, ellagitannins, or polyflavonoids whilst others produce a mix of all types of tannins.

In their natural state, hydrolysable tannins are a mixture of simple phenols with a low level of phenol substitution and low nucleophilicity. The total world production of commercial tannins is estimated at 220,000 tons per year, with a large percentage of condensed tannins (>90%) available on the market [5]. With high reactivity and a relatively low price, condensed tannins are both chemically and economically more interesting for the preparation of adhesives, resins, and gels.

1.2. Condensed Tannins

Condensed tannins are biosynthesized by the plant through their intermediate precursors (flavan-3-ols and flavan-3,4-diols) and other flavonoid analogs [6,7]. Their chemical structure is composed of flavonoids units that are subjected to various degrees of condensation. Traces of monoflavonoids or amino- and imino-acids are also reported in the composition of condensed tannins, but at too low concentrations to influence their chemical or physical properties [3]. However, other components in significant concentrations are often detected in tannin extracts which can modify the viscosity of the solutions. Among them, simple carbohydrates (hexoses, pentoses, and disaccharides) or carbohydrates chains of various length [8,9] can be linked to the flavonoid unit (Figure 1). Oligomers derived from hemicelluloses, complex glucuronates, and a low percentage of monoflavonoids (flavan-3,4-diols, flavan-3-ols, dihydroflavonoids, flavanones, chalcones, and coumaran-3-ols) could also be present in the extracts [6,7,10,11].

For example, in the black mimosa bark extract (*Acacia mearnsii*, formerly *mollissima*, de Wildt), it was reported 3–5% of monoflavonoids (flavan-3,4-diols and certain flavan-3-ols (catechin)) in its composition [7]. Thus, in this kind of tannin extract, each of the four combinations of resorcinol and phloroglucinol (A-rings) with catechol and pyrogallol (B-rings) coexist (Figure 2). In addition, the main polyphenolic pattern is represented by flavonoid analogs based on robinetinidin, and thus based on the resorcinol A-ring and pyrogallol B-ring. This pattern is reproduced in approximately 70% of the phenolic part of the tannin. The secondary but parallel pattern is based on fisetinidin, and thus on resorcinol A-rings and catechol B-rings. This represents about 25% of the total polyphenolic bark fraction. Superimposed on this two predominant patterns are two minor groups of A- and B-rings combinations.

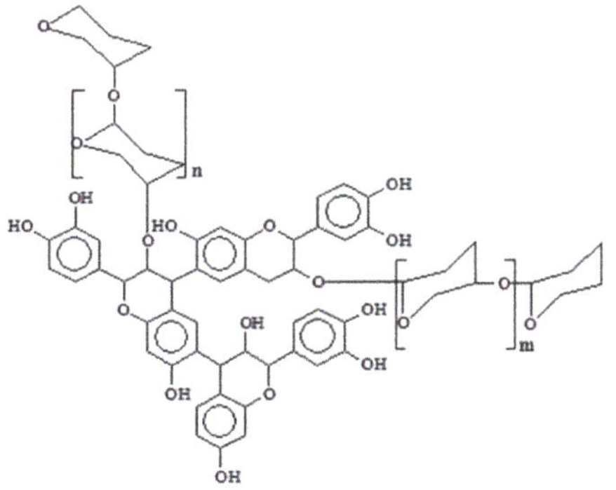

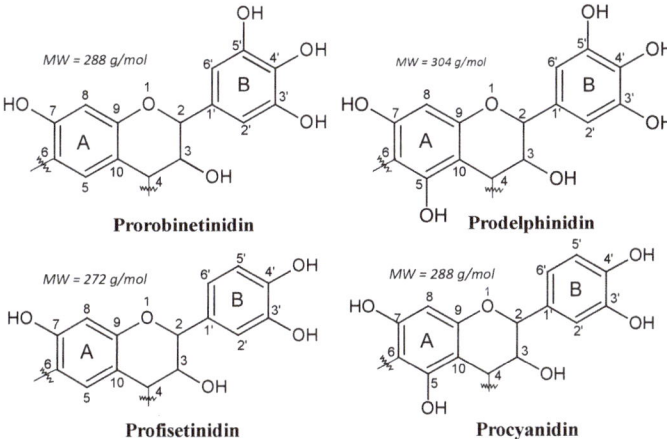

Figure 1. Polyflavonoids and carbohydrates linkage.

Figure 2. The four repeating flavonoid units in condensed tannins.

These are based on phloroglucinol (A-ring)-pyrogallol (B-ring) (gallocatechin/delphinidin) and on phloroglucinol (A-ring)-catechol (B-ring) flavonoids (catechin/epicatechin). These four patterns constitute 65%–84% of commercial mimosa bark extract. The remaining parts of mimosa bark extract are the so-called "non-tannins". This definition comes from the leather industry, where a "tannin" is considered to be any polyphenolic oligomer higher than, and comprising of, a trimer. It must be pointed out that the percentage of non-tannins varies considerably according to tannin extraction. The non-phenolic non-tannins can be subdivided into carbohydrates, hydrocolloid gums, and some amino and imino acid fractions [4,12].

1.3. Reactions of Condensed Flavonoid Tannins

Condensed flavonoid tannins are subjected to a number of reactions that impinge on their adaptability to different uses. These basic reactions in tannin chemistry have been abundantly described in the relevant review literature and the reader is recommended towards these reviews for more complete information [4,13,14]. The basic reactions of tannin are their rearrangements by:

1. Hydrolysis and acid or alkaline condensation [4,10,13,15–17]: This reaction leads to insoluble and unreactive compounds called "phlobaphenes" (Figure 3) or "tanners red" [18].
2. Sulfitation: This is one of the older reactions used in tannin chemistry to decrease the tannins viscosity in water and improve their water solubility [4,19], but the excess of sulfite can be deleterious for some applications [20].

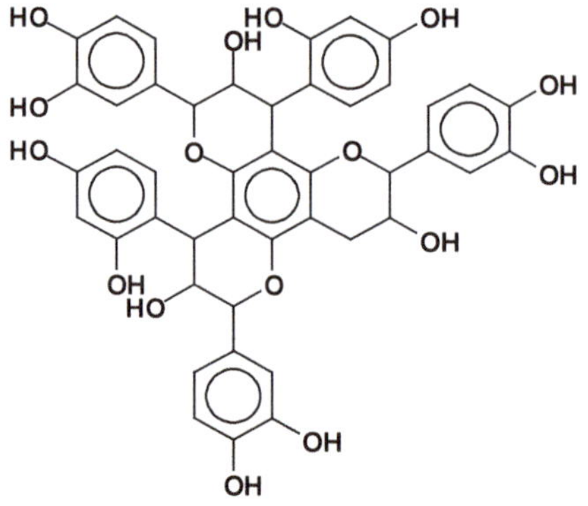

Figure 3. Chemical structure of phlobaphenes.

3. Catechinic acid rearrangement: While this rearrangement is easily shown to occur in model compounds where the reaction is carried out in solution [13,21], it is much less evident and easily avoidable in tannin extracts where the colloidal nature of the extract markedly limits its occurrence. This is fortunate, as otherwise some fast-reacting tannins such as pine, pecan, cube gambier, etc. could not be used to produce resins, adhesives, and other thermosetting plastics [16,22–26].

4. Catalytic tannin self-condensation: Polyflavonoid tannins have been found to self-condense and harden when in the presence of particular compounds acting as catalysts. Foremost, there is the catalytic effect of small amounts (2%–3%) of silica smoke, or nanosilica or silicates at high pH [27]. This reaction is fast and markedly exothermic, where a concentrated solution of tannin at 40%–50% in water gels and hardens at pH 12 and 25 °C in 20–30 min. The strong exothermic character of the reaction leads to this result, as the temperature increases by several tens of degrees in a short time [27]. Small amounts of boric acid and $AlCl_3$ were found to have the same effect but are much less exothermic.

5. Tannin complexation of metals: Tannins readily complex metal ions [28]. This characteristic is used to capture or precipitate toxic metals in water [29,30] and to isolate a rare metal such as germanium from its copper matrix, where it is mined for paint primers for metal application and several other applications. An old example is the formation of iron complexes, used to prepare intensely black/violet inks by the formation of ferric tannates (Figure 4). These coordination complexes are due to the ortho-diphenol hydroxyl groups on the tannin B-rings.

6. Reactions of tannins with formaldehyde and other aldehydes: Due to their phenolic nature, tannins undergo the same alkali- or acid-catalyzed reaction with formaldehyde as phenols (see more details in Section 2.2.1.).

7. Reactivity and orientation of electrophilic substitutions of flavonoids such as reaction with aldehydes: The relative accessibility and reactivity of flavonoid units is of interest for their use in resins, adhesives, and gels. A detailed description of reactions at different pHs is given in Section 2.2.2.

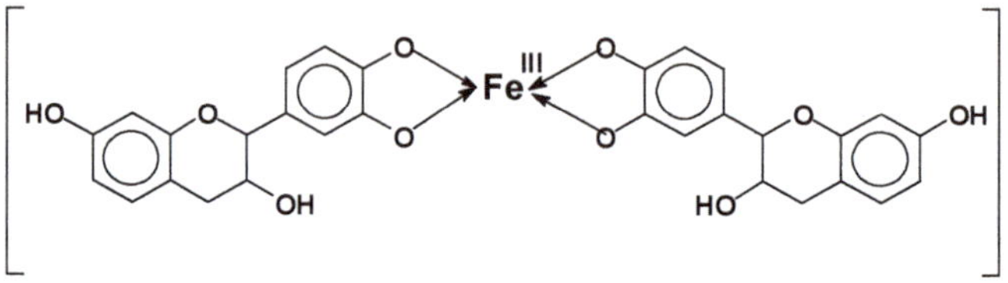

Figure 4. Ferric tannates [31].

While there is an abundant literature in chemical journals about tridimensional structure of flavonoid monomers, there is a scarcity of research in the literature about the three-dimensional spatial configuration of flavonoid oligomers. There is only one molecular mechanism study on this topic [32]. This study shows the correlation between the applicability of these materials and their 3D structure. For example, for a tetraflavonoid of 4,8-linked catechins, all 3,4-cis is in the helix configuration and when observing the helix axis, a characteristic structure presenting all four B-rings pointing outwards appears (Figure 5). Such a structure, rendering the hydroxyl groups of the B-rings particularly available, obviously facilitates their use and reactions, such as adhesion to a lignocellulosic substrate, formation of metallic coordination complexes [28–30], formation of polyurethanes with and without isocyanates [33,34], and other outcomes where the reaction of the B-ring is of interest [35], such as cross-linking at pH 10 and higher.

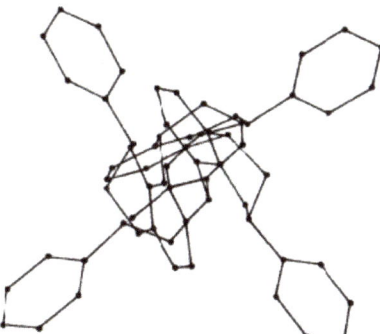

Figure 5. Tridimensional structure of a tetraflavonoid 4,8-linked.

Water solutions of 40%-50% polyflavonoid tannin extracts appear to be in a colloidal state as indicated by their zeta-potentials [36], this has been confirmed by ^{13}C NMR. This is caused by both the presence of noticeable proportions of hydrocolloid gums (fragments of hemicelluloses) as well as the presence of higher molecular mass tannins.

1.4. Industrial Extraction

Condensed flavonoid tannins are generally obtained from crushed bark or wood chips through countercurrent industrial hot water extraction (70–90 °C), but not under any pressure. Small percentages of sodium sulfite or metabisulfite can be added, sometimes with even lower proportions of sodium bicarbonate to improve solubility of oligomers of high molecular weight tannin. The sulfitation of tannins during their extraction, or after extraction, is one of the oldest and most useful reactions in the chemistry of condensed tannins. This process may be useful, but harmful if done excessively, depending on the end use for which the tannin extract is intended. Sulfitation allows the preparation of tannins of lower viscosity and increased solubility, which are therefore easier to handle [4,13]. After having passed through stainless steel tanks, a fine aqueous solution of tannin at very low concentration is first obtained, then concentrated to about 35% solids and spray-dried, or concentrated under vacuum to 86% concentration and bagged to form a solid mass, the so-called tannin extract "cast" [18].

The type of extraction process defined above has been used since the beginning of the 20th century for extracts of mimosa (*Acacia mearnsii*) and quebracho (*Schinopsis* sp.), which are the world's two largest sources of commercial condensed tannin extracts. The percentage yield of tannin extract is about 28–33% of the original weight of the bark or wood for these two species, which makes the extraction very profitable. Procyanidin or prodelphinidin tannins present generally higher molecular weight fractions and can undergo internal rearrangements, resulting in lower extraction yields of about 12–15%. Addition of a small percentage of urea to conventional solutions used for extraction has led to a higher yield level of 18–25%, and thus increased their economic interest [18]. Extraction by organic solvents, although giving higher yields, has proved to be expensive and unacceptable at the simple technological level of plant extraction. In addition, there are regularly percentages of other materials in the extracts, thus posing the problem of further purification of the tannin extract. In cases in which tannins are used for nutritional or pharmaceutical purposes, a second purification with organic solvents is necessary to remove the carbohydrates fraction present in the extract.

1.5. Industrial Uses of Condensed Tannins

There are several established industrial uses of condensed tannins, some of long date, some more recent. It is too long to explain in detail here, and the reader is invited to read relevant reviews on different subjects [31,37]. Only a few lines will be spent here to enumerate them in a very brief description. Traditionally, tannins have been used for (i) Leather tanning, the oldest and still the most important industrial applications, where tannin extracts have been used since the last quarter of the 19th century. However, tannin-containing bark had been used for this application since the second millennium before Christ [4]; (ii) Wood thermosetting adhesives, possibly the second largest use of tannin, but still far behind leather. This use takes up in part the slack of the leather market since the 1970s, but with new generations of wood adhesives still appearing [6]. Added adhesives for corrugated cardboards and for foundry sands must be used for this process [38]; (iii) Pharmaceutical applications, these being traditionally centered on troubles for the digestive system; (iv) Depressors of calcite ore flotation [8]; (v) Flocculants for the precipitation of heavy metal and other pollutants in water [11]; (vi) Stabilizers for oil and other deep level drilling; (vii) Additives for wine, beer and fruit juices; (viii) Ferric tannate writing inks, one of the oldest uses started from the 15th century [11]; (ix) Superplastifiers for cement [13]; and (x) metal antirust primer varnishes, this industrial usage that was in vogue in the 1950s and 1960s is now experiencing a revival of interest [4].

In addition to this more traditional long time operational, or even reasonably more recent usage, there is an increasing interest in researching new materials and applications for condensed tannins. At present, the interests that are more highly researched and focused on are (i) Fire resistant foams for several applications, such as thermal and acoustic insulation [39–41], as well as for hydroponic cultivations [19,42–44]; (ii) Hard thermosetting plastics for angle grinder disks and car brake pads [45,46]; (iii) New medical applications such as antiviral drugs and support for stem cells for bone reconstruction [8,47]; (iv) Polyurethanes with and without isocyanates for surface finishes and other applications [33,34]; (v) Adhesives for Teflon coatings on aluminum and steel for high temperature resistance [11]; (vi) Epoxy resins [48,49]; (vii) Flexible films [50] and flexible gels [51]; (viii) Fiber panels impregnated of tannin resins [18]; and (ix) Tannin-based gels, which will be extensively discussed in the next session.

2. Tannin Gels

2.1. Tannin Gels Synthesis

Organic porous tannin gels are versatile materials that can be used for several applications, mainly because of their final characteristics, which are normally tailored during synthesis steps. The tannin gel synthesis is based on a sol-gel process of soluble precursors. The colloidal suspension of solid particles (sol) is composed of tannin and a crosslinker dispersed in a solvent, commonly water. A schematic representation of a sol-gel process is given in Figure 6a.

Usually, the reactions start with an addition reaction between the main precursor (tannin) and the crosslinker (aldehyde), making these species more reactive. The partially hydrolyzed particles then crosslink by polycondensation reactions, releasing water molecules. Thus, the polymeric chains grow, forming a giant cluster of macroscopic size, becoming insoluble. This phenomenon is known as gelation [52]. The sol-gel transition is usually estimated by a significant change in viscosity through visual observation of the container tubes where the solutions are placed. Normally, reactions happen in a sealed flask at moderate temperature (50–85 °C), so that the whole solution can be converted into a gel [52], and when the viscous liquid is no longer flowing at an angle of 45°, the solution is considered gelled [53,54], establishing the gelling time (T_{gel}) [55]. The polymerization reaction in the aqueous medium generates a gel as a semi-solid system containing two phases: a solid based on a nanostructured network (spherical nodules) interconnected by narrow necks defined as "string-of-pearls", which is enclosed in a high porosity inter-penetrated solvent medium (second phase), as well as in some by-products formed during the polymerization reactions [52,56,57]. The size of these nodules depends directly on the synthesis conditions, especially the pH and the mass fraction of solids in the initial solution [58–61]. Usually, phenolic gels prepared under acid conditions lead to a tridimensional chain with large nodules, consequently developing larger pores and higher pore volumes [60–63]. By increasing the pH to an alkaline level, a less porous material is created, with smaller pores or non-porosity [60,61,64]. Since tannin gels are porous materials from a phenolic precursor, they also follow this trend, as presented in Figure 6b,c.

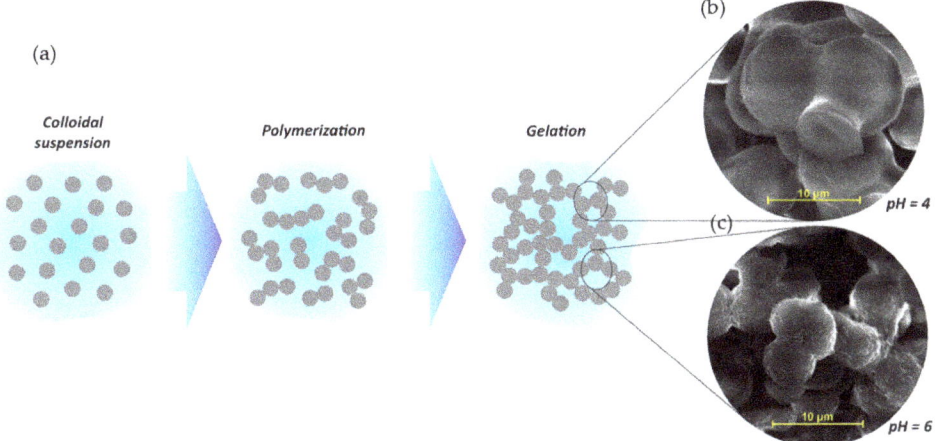

Figure 6. (a) Schematic representation of sol-gel process of tannin gels and their respective SEM images at pH 4 (b) and pH 6 (c). Adapted with permission from reference [61]. Copyright 2013 Elsevier.

After gelation, the polymerization reactions still continue, since the network formed is highly flexible and its constituent chains can move in relation to each other [62]. At this point, small clusters still exist and covalent bonds cross-linking occurs within the main network [52,55,62]. Besides, a fraction of the water in the polymerized gel structure is present as methylol derivatives as –OH groups [62], and part of the porosity is still prone to evolve, since evaporation of water generates voids within the network. Therefore, gel drying should not be done right after gelling because an ageing step is required to ensure the maximum formation of crosslinks [52,62,65]. These additional reactions generate a porous material with better mechanical properties, i.e., are able to withstand capillary forces during the drying process [62].

The tannin gels formed are classified as physical or chemical, according to the type of crosslinks in their network structure [60]. The so-called physical gels are formed by weak bonds, usually based on Van der Waals and other secondary forces [66] such as hydrogen bonds [67]. Chemical gels instead have a structure based on strong covalent bonds crosslinks, establishing a reticulated network and an infused gel [67].

2.2. Mechanisms

2.2.1. Tannin-Formaldehyde-Systems

The reactions of tannin gels are based on polymerization of flavonoids units from condensed tannins with formaldehyde (the most used aldehyde for the production of tannin gels), especially with the flavonoids A-rings through methylene bridge linkages [14,68,69]. However, experimental studies suggest that the B-rings (catechol or pyrogallol) are also able to react with formaldehyde in a more reactive medium, with more acidic or alkaline [70], or with the addition of zinc acetate [71–75]. The main cross-linking reactions involving tannin-formaldehyde systems are the formation of both methylene bridges ($-CH_2-$) and unstable methylene ether bridges ($-CH_2OCH_2-$) as shown in Figure 7. The latter is unstable and thus easily rearranged, forming methylene bridges and releasing formaldehyde [14]. Additionally, carbohydrates and complex glucuronates present in tannin extracts also react with formaldehyde, which might assist in network formation [14]. During the formation of the tannin-formaldehyde network, the initial immobilization of the network places far from each other a number of potentially reactive sites thus preventing the formation of further methylene bridges. Substitution of formaldehyde by other aldehydes, by up to 30%, improves the resin cure. The reactions of tannins with aldehydes that have led to their extensive applications as wood adhesives will not be discussed here, since extensive reviews can be found elsewhere [14,42,70].

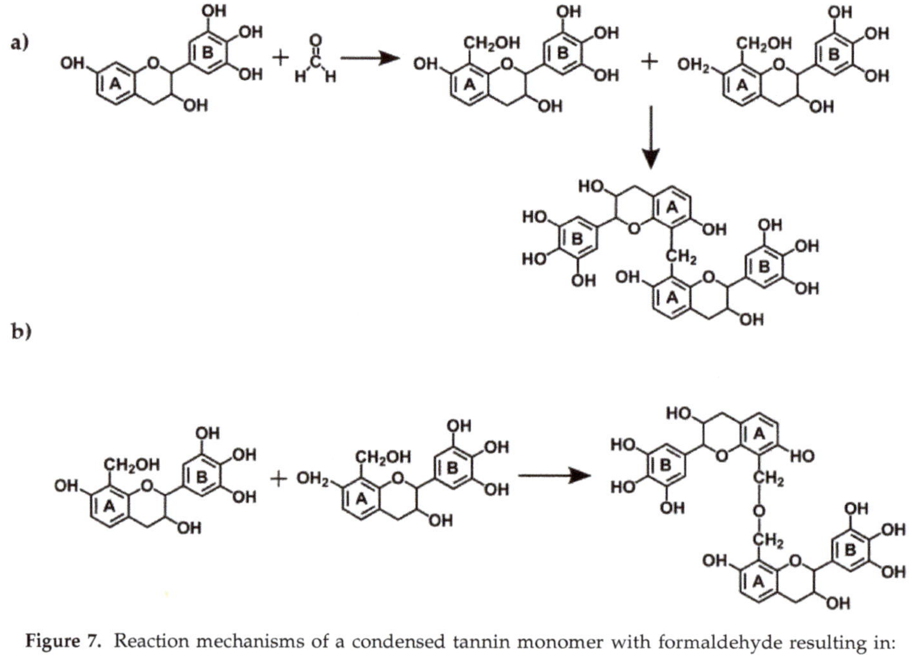

Figure 7. Reaction mechanisms of a condensed tannin monomer with formaldehyde resulting in: (a) methylene bridges and (b) methyne ether bridges.

2.2.2. Influence of pH and Mass Fraction

As mentioned before, pH control is extremely important in the reactional system of phenolics. The C8 site on the A-ring (Figure 2) is the first one to react, e.g., with an aldehyde, and when free, it is the site with higher reactivity [10,70]. The C6 site on the A-ring is also very reactive, but less than the C8 site, since this latter presents lower steric hindrance [10,70]. Generally, the reactions involve only these two sites on the A-ring. The B-ring is particularly unreactive. A low degree of substitution at the 6' site of the B-ring can occur (Figure 8). In general, at higher pHs such as pH 10, the B-ring starts to react too, contributing to cross-linking as well [68,76]. Thus, for catechins and phloroguinol A-ring type flavonoids, the reactivity sequence of sites is C8 > C6 > C6' when these are free. For robinetinidin and fisetinidin, thus for resorcinol A-ring type flavonoids, the reactivity sequence is modified to C6 > C8 > C6' due to the greater accessibility and lower possibility of steric hindrance of the C6 site (Figure 8) [10,68].

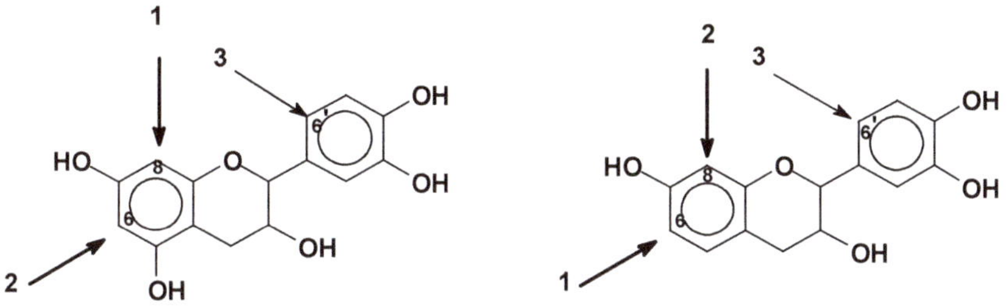

Figure 8. Reactive sites of flavonoid units.

The T_{gel} is a useful parameter to confirm the precursor reactivity at different pHs. The curve of T_{gel} of flavonoid tannins with aldehydes has always the shape of a bell curve. Normally, the longest T_{gel} is around pH 4, while the fastest ones are at lower and higher pHs, depending on the tannin system studied. The curve reaches an almost asymptotic plateau of very high reactivity and short T_{gel} around pH ≥ 10 and at pHs < 1–2 [42,70]. The maximum T_{gel}, i.e., the reaction at the lowest condensation rate, changes as a function of both the temperature and the reaction system. Thus, tannin-formaldehyde gel has the highest T_{gel} at pH 4–5 (50 °C) [77].

During their polymerization reactions, the tannin molecules may become inaccessible to reagents due to the network early immobilization and tridimensional structure, as mentioned before. Such a steric hindrance is associated with a lack of flexibility, especially when the network tannin-formaldehyde forms already at low degree of condensation. In this case, the prospective residual reactive sites become too distant to participate to the cross-linking leading to incomplete polymerization [13].

Mass fractions of total solids from initial solutions (Equation (1)) also represent an important parameter in the production of tannin gels. In the system tannin-formaldehyde, for example, low concentrations of mass fraction (<18 wt.%) induce gels with high pore volumes. However, the low final density results in a fragile structure, subjected to a higher shrinkage during the drying process [60] and partial porosity losses. Increasing the mass fraction of solids from the initial solution (22–40 wt.%) tends to produce gels more consistent, but with less porosity [60]. It is important to notice that there is always an optimal condition depending on the final textural properties required. Thus, the experimental parameters as pH and mass fraction should be optimized and chosen to fit the desired application.

$$Mass\ fraction\ (\%) = \frac{m_{solids}}{m_{solids} + m_{solvent}} \tag{1}$$

2.2.3. Tannin-Soy-Formaldehyde Gels

Organic gels can be produced using only tannin as the raw biopolymer, or this can be combined with a second biosourced material, such as lignin [78] or protein from soy flour [54]. Soy-tannin-formaldehyde is known as an adhesive for wood particleboard [79]. Other natural materials such as albumin have been already applied for the production of a macroporous monolith (tannin-albumin-formaldehyde) [80]. At the same time, diluted adhesives might be explored in the preparation of gels [81,82]. A reaction mechanism was proposed for tannin-soy-formaldehyde gels (Figure 9a) [54]. The authors showed the main reactions occurring using FTIR, ^{13}C-NMR, and XPS analyses. First, the soy protein is denatured to expose its amide groups (N), followed by an addition reaction with formaldehyde, and finally by a copolymerization with the tannin (T). The main cross-linking reactions occur between tannin-soy

(N-CH$_2$-T) or soy-soy (N-CH$_2$-N) and tannin-tannin (T-CH$_2$-T) through methylene bridge linkages, as demonstrated by ^{13}C-NMR spectra (Figure 9b). The reactivity of tannin-soy-formaldehyde gels slightly changes, so its highest T$_{gel}$ was found to be at pH 6 and 85 °C [54] compared to pH 4–5 and 50 °C for tannin-resorcinol-formaldehyde gels [53].

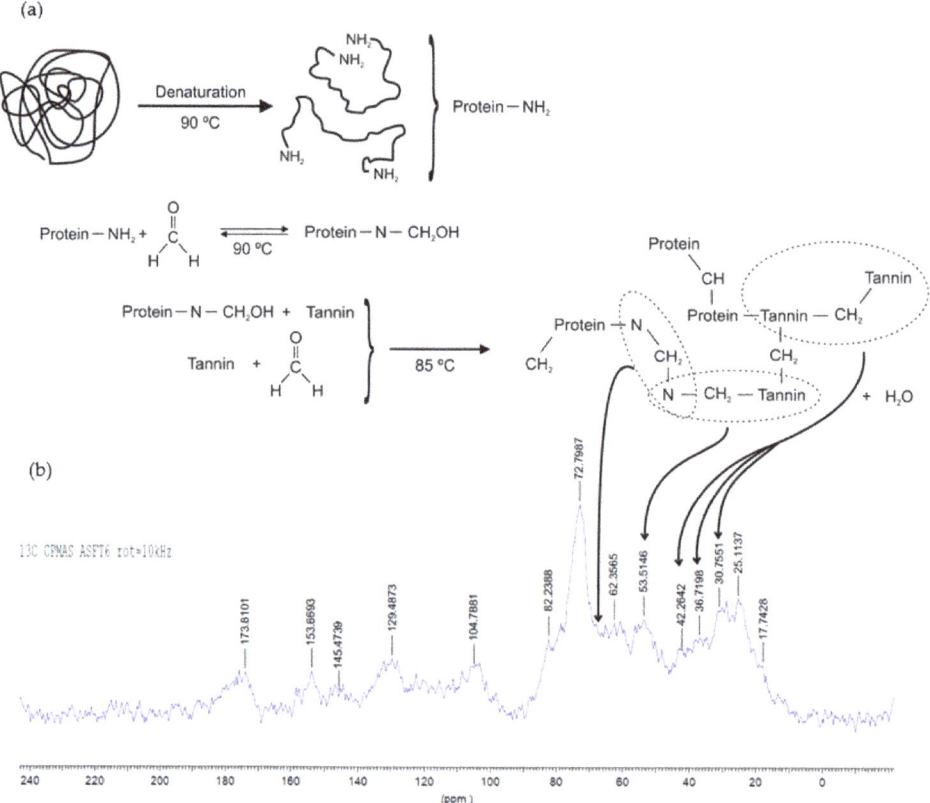

Figure 9. Suggested cross-linking reactions on tannin-soy-formaldehyde gel (**a**) and ^{13}C-NMR spectra of organic gel at pH 6 (**b**) [54].

2.3. Preparation Methods and Conditions

2.3.1. Hydrogels Formulations at Normal Conditions

After dissolution of all reactants, the final solution is usually placed in a hermetically sealed container to avoid evaporation of liquid, and then kept at determined temperatures (50–85 °C) over five days for gelation and ageing. The latter conditions used during the preparation of tannin gels will be considered in this review as the "normal conditions". After this period, materials were left to cool down at room temperature before the solvent exchange step. It is worth noting that materials prepared with no ageing or temperature steps are also referred as tannin gels in the literature. Normally, the authors synthesized gels by mixing tannin with formaldehyde, leaving them for several hours at room temperature, and drying them at temperatures higher than 45 °C to generate a final product with a gummy aspect [83–88]. However, those gels are much less porous (e.g., up to 6 m^2/g [88]) but largely employed as adsorbents for organic and inorganic contaminants in water treatment at a laboratory scale (see Section 2.6). Thus, these materials will be termed as *tannin-gel* in this review to differentiate the highly porous gels prepared after the gelation, ageing, and drying steps.

The majority of tannin gels prepared in presence of an aldehyde at a different mass fraction, mass ratio, and pH are considered as tannin chemical gels. However, physical tannin-based gels can also be obtained by varying the mass fraction and the pH in the following proportions, e.g., 4 wt.% at pH 6 and 10 wt.% at pH 8. In such conditions, the solutions are gelled only after been removed from the oven (after cooling down) but they become a liquid again when they are returned to the oven (melting point around 85 °C). This phenomenon is not been fully understood [60].

2.3.2. Solvent Exchange

The tannin-based gel so prepared is named after the solvent employed during its synthesis. Hydrogel or aquagel fits for water, while alcogel is applied for gels prepared in alcohol medium [89]. The formed solid is an organic macromolecule, which has saturated pores with (i) Water and/or solvent; (ii) Unreacted residual products; and possibly (iii) By-products of the polycondensation reactions. The liquid in the pores must be replaced by air through a non-destructive procedure, ensuring that the formed nanostructure is not destroyed [90] and most of the porosity is preserved [91].

To do so, an intermediate procedure is required. The produced hydrogel is subjected to a solvent exchange, i.e., the liquid present in the porosity (water, alcohol, by-products) is replaced by an appropriate solvent accordingly to the desired drying method. Solvents such as acetone [77], ethanol/carbon dioxide [60], or *tert*-butanol [53] are usually employed. In the literature, most authors exchange their hydrogels for three days, replacing the solvent every day with a fresh one and providing an effective solvent exchange within the tridimensional structure of the gel [60,77].

2.3.3. Drying Conditions and Final Physicochemical Properties of Tannin Gels

After solvent exchange, tannin gels are finally ready to be dried. There are three types of drying procedures that are commonly used in the production of gels, which result in different porous materials with distinct textural properties:

1. Subcritical drying: Gels are dried under atmospheric conditions to form xerogels.
2. Freeze-drying: Gels are dried at freezing conditions to produce cryogels.
3. Supercritical drying: Gels are dried at a critical point of a working fluid to produce aerogels.

The respective dried materials from a gel prepared from tannin and formaldehyde with a resin mass fraction of 6 wt.% and an initial pH of 2 (aerogel, cryogel and xerogel) are presented in Figure 10 [92]. Visually, it is possible to notice macro differences: (i) Xerogel presents a considerable shrinkage; (ii) Cryogel shows micro cracks due to the formation of ice crystals coming from solvent during the freezing stage; and (iii) Aerogel presents a better preservation of the initial volume and porosity. More details about each of these methods are described below.

Subcritical drying allows the evaporation of solvent at room temperature, or using an oven at temperatures up to 50 °C. It is the simplest and the cheapest method of gels drying. However, the disadvantage of this technique is related to the formation of a liquid meniscus in each pore while solvent evaporates from the surface of the gel. The capillary forces induced by the solvent within the pores generate pressures differences between 100 and 200 MPa [93], which cause an extreme decrease in the final material porosity.

Lyophilization is a drying process based on freezing the solvent present in the pores, followed by its sublimation [94]. However, during solvent freezing, dimensional variations of solvent occur, causing tensions in the gel structure. This can cause fissures or even lead to a complete destruction of the initial geometric gel structure, resulting in a powder as a final product [63]. Therefore, the use of a solvent that has minimum volume variation during freezing is required, coupled with a high vapor pressure to promote sublimation. *Tert*-butanol (2-methyl-2-propanol) is generally used to minimize the effects of volume and structural modification of a cryogel due to its low-density (-3.4×10^{-4} g/cm^3) and low vapor pressure variation (821 Pa) [94] at the freezing point compared to water, -7.5×10^{-2} g/cm^3 and 61 Pa, respectively [95].

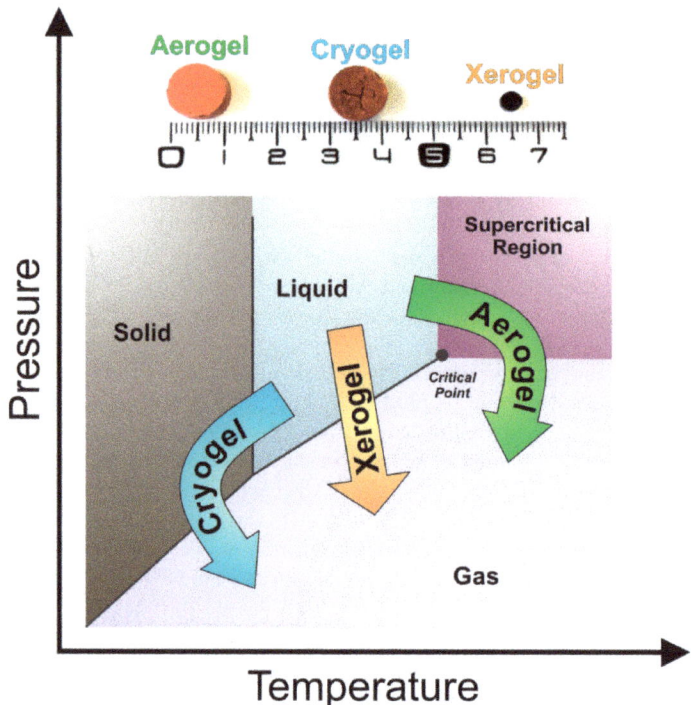

Figure 10. Phase diagram of the solvent within the gel structure and the representation of the different drying methods with their respective porous materials, adapted with permission from reference [92].

The supercritical drying technique is based on increasing both the pressure and temperature of the solvent beyond the critical point to avoid the formation of a vapor-liquid meniscus in the hydrogel pores. Such a technique minimizes gel shrinkage, and consequently the porosity loss due to low capillary forces generated [90,96]. Organic solvents such as acetone [77] are used for drying tannin based gels to produce aerogels at a critical temperature and pressure of 250 °C and 14 MPa, respectively. This is known as the "HOT process" where the drying step is carried out at high temperature conditions [66]. It can also be performed in the presence of CO_2 [60], which is called the "COLD process" [65,90] at a critical temperature and pressure of 40 °C and 10.4 MPa, respectively. The latter requires the exchange of two solvents due to low solubility of CO_2 in water. First, water is replaced by ethanol, followed by liquid CO_2 exchanging.

Usually, aerogels maintain large part of their geometric and nanometric structures, which is associated to lower volume shrinkage. Thus, the initial porosity is largely preserved, and aerogels regularly present low values of bulk density and high values of specific surface area and pore volumes [60,77,89].

In order to avoid capillary tension, gels can be synthesized directly in solvents with surface tension lower than water, such as acetone or ethanol. However, as water is always produced during polycondensation reactions [62], the formation of a vapor-liquid meniscus could not be totally avoided. Thus, surfactants are employed in gels synthesis to reduce the effects of surface tension in xerogels during their drying [61,97,98] (see more details in Table 1). Furthermore, surfactants may also be used as a template to produce ordered porous materials based in a self-assembled micellar system [99]. Xerogels prepared from tannin-formaldehyde with a mass fraction of 25 wt.% and surfactant (Pluronic F-127) (Figure 11), had bulk densities about (0.28–0.65) g/cm^3 comparable of tannin aerogels [98]. The numbers 2 to 10 in Figure 11 refer to initial pH of tannin-formaldehyde-pluronic solutions, and their highest T$_{gel}$ (~240 min) was found to be at pH 4 and 85 °C [61].

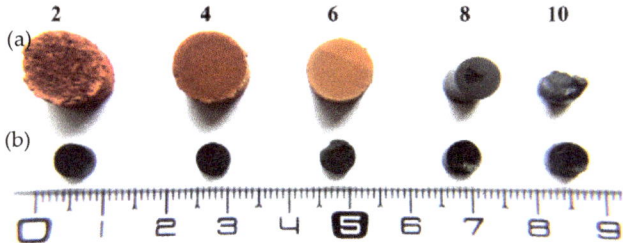

Figure 11. Tannin-formaldehyde gels prepared with surfactant (**a**) top view, and without (**b**) bottom view. Reprinted with permission from reference [98]. Copyright 2011 Elsevier.

The use of additives, e.g., surfactants, can decrease the capillary stresses during the drying step. Tannin can indeed be used in a wide range of pH (2–10), differently from lignin and phenol (that only work at alkali pHs) or resorcinol (that only works from mildly acid to alkali pHs). The final texture properties of organic porous gels change with several parameters, such as the pH of the initial solution, the mass fraction of solids, the raw materials, the chemicals employed to modify the initial pH of the solution (alkali and acids), the temperature of gelation, and the drying method used. Thus, there is no unique recipe to control the final properties of gels, since each parameter plays a specific role, as explained before. There is an optimal condition for each system, depending on the desired final property.

A summarized description of organic tannin gels prepared with formaldehyde by different systems, as well as the description of the drying methods used and their physicochemical properties, are reported in Table 1. To produce an organic gel with high specific surface area, it is preferable to choose materials with the highest gelation time (which depends on the pH) and with a mass fraction between intermediate to more diluted (<20 wt. %). High mesopore volumes are used for tannin gels prepared in the presence of soy, whereas monolithic, unimodal, or bimodal organic gels are readily prepared in the presence of a surfactant such as pluronic.

2.4. Hydrogels Formulations under Hydrothermal Conditions

Hydrothermal carbonization (HTC) is a wet thermochemical conversion process to convert biomass (or wet biomass) into fuels in solid and liquid phases [100]. This technique has been known since 1913, but there has been an increasing interest in it only during the last decade. HTC uses moderate temperature (180–260 °C) and pressure (1–4.7 MPa) to transform biomass precursors into hydrochars with a hydrophilic shell and a hydrophobic core. Recently, the HTC technique has allowed the production of gels made from various precursors, e.g., phloroglucinol with some carbohydrates (glucose, fructose, or xylose) [101], borax and glucose [102], S-(2-thienyl)-l-cysteine and 2-thienyl carboxaldehyde [103], D-glucose and ovalbumin [104] and mimosa tannin in ammonia solution [105], without any crosslinker agent (formaldehyde) and at a much lower reaction time (e.g., 24 h).

Figure 12 presents the different organic gels prepared under HTC conditions with: (a) tannin aqueous solution at low pH [106]; (b) tannin aqueous solution in presence of a metal salt; and (c) evaporated aminated tannin. The first gels (Figure 12 a) were elaborated under HTC conditions (180 °C for 24 h) simply with tannin solution in a very acidic environment (pH lower than 3). Three main reactions occurred with acidic tannin solution under heat: (i) the formation of soluble anthocyanidin and catechin [70,107]; (ii) the rearrangement to phlobaphenes, which are insoluble in water and have high molecular weight (Figure 2) [18,108] and; (iii) possibly condensation between the free radical coupling of B-ring catechol units in the presence of atmospheric oxygen [70]. In the same HTC conditions, tannin dissolved with metal salts, especially chromium, lead to a nano-structure powder material, but its microscope structure showed small particles aggregated together in a nodular form, which is typical of a gel structure [94]. This is due to the tannin characteristic of being good metal chelators (Figure 4). Thus, metal ions were chelated with most of hydroxyl groups in tannin-based polymer networks under HTC conditions (Figure 12b).

Another gel produced under the same hydrothermal conditions as previously mentioned was prepared from aminated tannin. In a typical synthesis, tannin was dissolved in ammonia solution at 28 wt.% and left in a fume hood to evaporate overnight. The solid dried residue recovered was then mixed with water. The final mixture after HTC conditions generated a monolithic material, having the same typical structure of a gel made from tannin and formaldehyde, but in absence of any crosslinker or condensation catalyst [105,109]. The material is composed of very long chains polymerized to an "infinite" three-dimensional network that were not even detected by MALDI-ToF analysis due to their high molecular weight. The mechanism of gel formation is based on selfcondensation and partly dehydrated tannin, mainly through the heterocycle opening [105,109]. The same reactions were also reported during tannin amination under ambient conditions [110]. However, under HTC conditions, the formation of this gel is probably due to the creation of −NH− bridges with the amination of C3' and C4' (Figure 2) according to ^{13}C NMR analysis, as shown in Figure 12c.

The evaporated aminated tannin gel was also dried accordingly to produce xerogel, cryogel and aerogel at the same conditions as mentioned before with tannin-formaldehyde gels [111]. After HTC, these gels were homogeneous and had monolithic shapes, but they were quite fragile, with some cracks here and there. The mass fraction of evaporated aminated tannin was varied at 11, 18, and 27 wt.% to check any possible physicochemical differences with more concentrated gels. Interestingly, xerogels had great porosity, and the drying method had an important impact on the most diluted gel (11 wt.%). At such concentration, the porosity was improved in this sequence: xerogel, cryogel and aerogel (102, 219, and 295 m^2/g, respectively). The same trend was also observed for tannin-formaldehyde gels under normal conditions (Figure 10).

A hydrothermal cryogel made from graphene oxide sheets and tannin was developed by Deng et al. [112]. The mechanism of gel formation was based on the self-assembly of graphene oxide with tannin acting as both reductant and surface functionalization agent. The final monolith with a three-dimensional structure had a mesoporous structure and a good performance for strontium removal due to its important amount of oxygen functional groups (see more details in Section 2.6). A summarized description of organic tannin gels made through HTC synthesis conditions and their final physicochemical properties can be seen in Table 2.

Biomolecules **2019**, *9*, 587

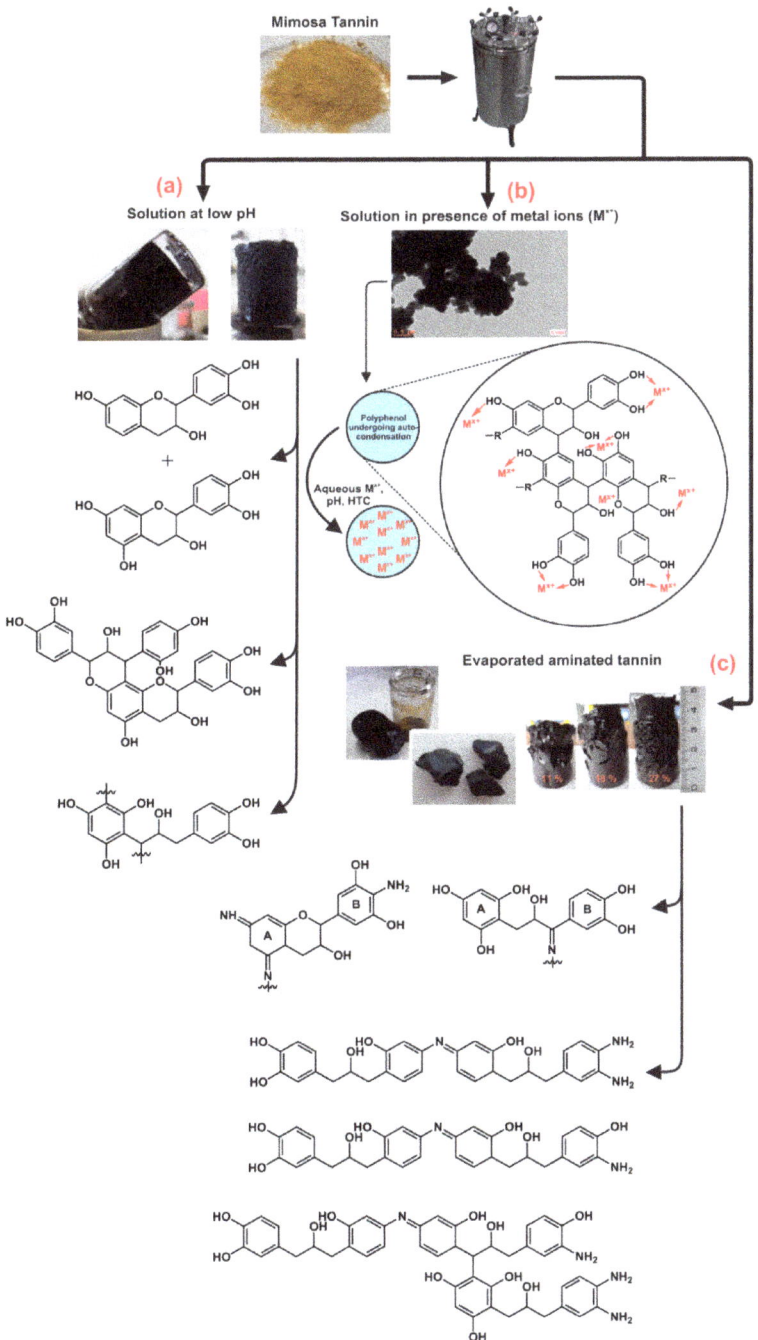

Figure 12. Gels prepared under hydrothermal conditions with aqueous tannin solution at low pH (**a**) (Reprinted with permission from reference [106]. Copyright 2015 Elsevier), with metal salts (**b**) [113], and with aqueous evaporated aminated tannin (**c**) [114].

2.5. Thermal Treatment of Organic Gels

Thermal treatments are very well-known techniques used to transform organic-derived materials into carbonaceous materials from different precursors. Typically, carbonization and activation are the main methods used for the synthesis of carbon gels. Carbonization is based on heating up a dried organic material in an inert atmosphere (nitrogen, helium or argon) at temperatures between 600 and 1000 °C. At such conditions and at slow heating rates, the volatiles are released in a certain form to avoid the shrinkage of the final material structure [115]. Although it may vary from one precursor to another, the influence of temperature, residence time and heating rate have been well studied. For example, it has been demonstrated that temperatures above 900 °C are suitable for removing most volatile compounds [77], developing the porosity of the tannin gel, improving its micropore volume and surface area, and widening the narrow pores as well. This was indeed confirmed by Szczurek et al. [77] through FTIR analysis of the tannin-formaldehyde gel structure after drying and carbonizing at different temperatures (300–900 °C). At lower temperatures (300–600 °C), the functional groups (mostly oxygenated) were still present in their chemical structure, whereas at higher temperatures (700–900 °C), the same groups were drastically reduced due to their gasification during carbonization. Thus, temperature has a positive influence on the formation of microporosity, surface area and fixed C parameters, but it has a negative influence on carbon yield, average porosity, and functional groups that will be volatilized (e.g., N, H, and O content).

For specific applications (e.g., energy storage systems), gel materials could have very high porosity and surface area, reaching up to 3000 m^2/g [116]. Thus, another largely known method for improving the textural properties is the activation method. Different organic precursors can be submitted to an activation process (physical or chemical) before or after carbonization. In physical activation, organic gels are introduced in an oven in an inert atmosphere, but also in the presence of CO_2 or steam at temperatures between 700 and 900 °C [117]. In chemical activation, the organic gel is in contact with a chemical agent instead such as H_3PO_4, KOH, NaOH, $ZnCl_2$, etc. at temperatures between 400 and 800 °C, which are considerably below the physical activation temperatures [118]. However, the disadvantage of this process is that the final material must be washed to remove residual chemical agents and possibly inorganic materials formed during this kind of activation [119]. The carbonized or activated carbons have higher specific surface area, higher pore volume, resistance against being attacked in acid and basic solutions, hardness, ignition at low temperature (200–500 °C), hydrophobicity and higher electrical conductivity which can be applied for catalysis, gas/air purification, and in the mining, chemicals, water, food, and pharmaceutical industries [120,121].

As mentioned before, carbon gels prepared under normal or hydrothermal conditions present textural properties tunable by pH, mass fraction, drying conditions, and others. In the case of gels prepared under normal conditions in presence of formaldehyde, the influence of the pH follows the trend: larger nodules for gels prepared under acid pHs and smaller nodules at alkali pHs (Figure 13a,b). In relation to their textural properties, Table 1 summarizes the main findings. In brief, tannin-formaldehyde gels after carbonization or activation reached the highest values of surface areas (S_{BET}) of 1420 m^2/g and 1810 m^2/g, respectively. In contrast, the highest S_{BET} values for organic gels were attained at 880 m^2/g. Thus, the thermal treatment is crucial on the development of porosity, and consequently on the surface area. In addition, these values are comparable to phenol-formaldehyde [122] and resorcinol-formaldehyde [63] aerogels (up to 520 m^2/g and 695 m^2/g, respectively) and their carbon derivatives (up to 720 m^2/g and 580 m^2/g, respectively).

Table 2 also summarizes the description of tannin carbon gels preparation and their physicochemical and textural properties under HTC conditions. Tannin carbon gels made with different pHs under such conditions had a significant influence on the reaction rate, carbon yield and on the physicochemical characteristics of the final products. Firstly, by modifying the reaction medium, the hydrochar yield increased from 65 wt.% (HTC of tannin solution at non modified pH of 4.2) to 87 wt.% (HTC of tannin solution at pH 1) [106]. The low pH had a positive impact on the final carbon yield and on the lower nodule diameter (Figure 13c,d). The pH of tannin gels made under hydrothermal conditions somehow did not follow the same tendency as for tannin-formaldehyde gels. The gel structure with a lower nodule diameter (at low pH) might have promoted the evolution of volatiles, improving the development of its surface area with almost the same microporous size distribution (800 m^2/g and 91% micropores) when compared to that with no pH modification (600 m^2/g and 96% of micropores) [106].

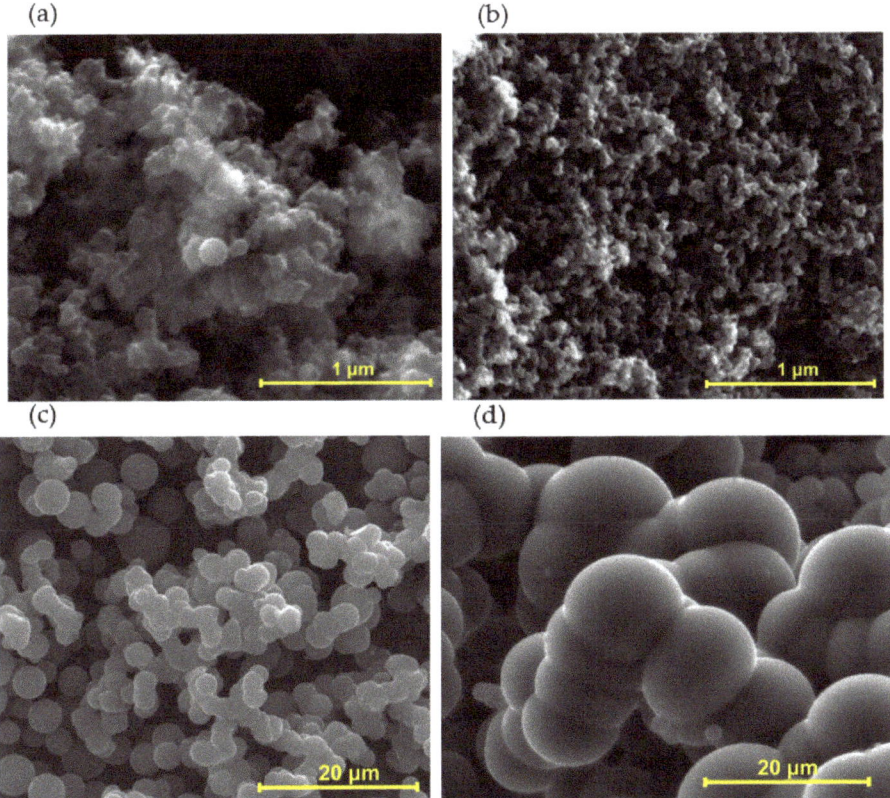

Figure 13. Tannin-formaldehyde carbon aerogels prepared under normal conditions at pH 3.3 (**a**) and pH 8.3 (**b**) (Reprinted with permission from reference [77] Copyright 2011 Elsevier); and hydrothermal carbon gels prepared with tannin solution at pH 2 (**c**) (Reprinted with permission from reference [106] Copyright 2015 Elsevier) and at non-modified pH (4.2) (**d**) [114].

Carbon aero-, cryo- and xerogels made from HTC conditions had also similar physicochemical characteristics of gels made from phenol/resorcinol-formaldehyde carbon aerogels [122,123] previously discussed. For example, surface areas of an aerogel made from HTC of evaporated aminated tannin at 27 wt.% and posterior carbonization at 900 °C reached values of up to 900 m^2/g, having a mixture of well-developed microporosity (54%) as well as wider microporosity and mesoporosity (46%). However, these gels made under HTC were not considered to be monolithic compared to tannin-formaldehyde gel materials. Thus, possible applications for such gels would be in electrochemistry, as the material must be crushed and pressed during the preparation of carbon gel electrodes (see Table 3).

Table 1. Organic and carbon gels synthesized at normal conditions.

Precursors	Conditions	Main Findings	Ref.
		Hydrogels Prepared Under Normal Conditions	
		Organic Tannin Gels	
Tannin + formaldehyde (TF)	Hydrogels prepared at 85 °C for 120 h; Mass fraction of total solids (MFTS) of 4–40 wt.%; NaOH and p-toluene sulfonic acid were used to change the initial pH of the solution (4.3); pH = 2–10; Supercritical drying in CO_2	Organic aerogels; S_{BET} = 219–880 m^2/g; Porosity = 40–97% The highest S_{BET} (880 m^2/g) and the highest micropore volume (0.28 cm^3/g) were obtained for a gel prepared with MFTS of 10 wt.% and pH 6. The maximum mesopore volume (1.34 cm^3/g) and the highest pore volume (21.5 cm^3/g) were found for samples with a MFTS of 18 and 4 wt.% at pH 8 and 2, respectively. The transparency of gels was associated with the structure of the aquagel: larger nodules at low MFTS (< 18 wt.%) and low pH (especially 2) lead to larger clusters and pore volumes (21.5–4.9 cm^3/g), reducing the transparency of the final gel. By increasing the pH (4–10) and MFTS (22–40 wt.%), the nodules were reduced, and the transparency tended to increase. Aerogels with MFTS greater than 22 wt.% with extremely high pH (10) resulted in a low porous material (0.48–1.01 cm^3/g). The most diluted samples (4 and 6% of TF resin at pH 6 and 8, respectively) presented the highest shrinkage volumes (~87%) during drying step, which was caused by high capillary forces due to their low mechanical properties. These values are comparable to those of aerogels from natural biopolymer, such as cellulose (75 wt.%). The S_{BET} values found for samples with MFTS lower than 20 wt.% were higher than those of any other bioresourced organic aerogel.	[60]
Tannin + formaldehyde + Pluronic (TFP)	Hydrogels prepared at 85 °C for 120 h with Pluronic® F-127, using a mass ratio of T/P of 2; NaOH and p-toluene sulfonic acid were used to change the initial pH of the solution (4.9); pH = 2–10; Subcritical drying in two steps: air and 80 °C	Organic xerogels; S_{BET} = < 1.5 m^2/g; Porosity = 2–79% The surfactant P was employed as an additive to decrease the surface tension of the aqueous system inner porosity and consequently mitigate the shrinkage during drying step. The system TFP presented a correlation between porosity characteristics (pore volume and centered peak pore size measured through mercury porosimetry) and pH. Thus, pH played a crucial role in pluronic–tannin gels preparation. The xerogels prepared at pH 2–7 presented a bulk density comparable to those of aerogels (0.30–0.58 cm^3/g). Such a system demonstrated an interesting and low-cost technique for preparing porous gels with tailored porosity on a large scale. The final xerogels presented unimodal (macroporosity) or bimodal (macro-mesoporosity) porosity.	[61]
Tannin + soy + formaldehyde (TSF)	Hydrogels prepared at 85 °C for 120 h; Proportions of T/S resin of 30 and 70 wt.% (on a dry basis), respectively; NaOH in pellets and H_3PO_4 (85%) were used to change the initial pH of the solution (8.2); pH 5.5–9; Mass ratio of total solids 14 wt.%; Supercritical drying in CO_2	Organic aerogels; S_{BET} = 384–478 m^2/g; Porosity = 84–88% The pH of the initial solution had a significant impact on gelation: the lowest gelation time, which happened at pH 6, produced the highest S_{BET} (478 m^2/g) and micropore volume (0.15 cm^3/g). Curiously, this aerogel also presented the highest mesopore volume (2.29 cm^3/g). A correlation between S_{BET} and gelation time was noticed. The TSF aerogels were based on a filamentous structure, which is comparable to cellulosic and lignocellulosic aerogels. Organic aerogels are natural at 91%, and mesopore volumes (1.72–2.29 cm^3/g) are among the greatest ever volumes reported for aerogels of natural origin, considering the same density (0.19–0.25 g/cm^3).	[54]
Tannin + lignin + formaldehyde (TLF)	Hydrogels prepared at 85 °C for 120 h; Mass ratio of T/L of 0.11–1 (on a dry basis); Mass ratio of (L + T)/F) of 0.83–2.5 (on a dry basis); Initial pH of 10; Supercritical drying in CO_2	Organic aerogels; S_{BET} = 50–550 m^2/g; Porosity = 72–87% The S_{BET} of TLF aerogels was mostly related to mesopores volumes (0.2–1.4 g/cm^3), since micropores were extremely low (0.01–0.02 cm^3/g). The highest S_{BET} values were found for TLF gels at mass fractions of T/L = 1 and (L + T)/F = 1.25, demonstrating that tannin could be replaced by lignin at 50 wt.%. Besides, there is an optimal quantity of formaldehyde that can be used to produce highly porous aerogels. Organic aerogels were synthesized using up to 71% of natural precursor.	[78]

Table 1. *Cont.*

Precursors	Conditions	Main Findings	Ref.
Hydrogels Prepared Under Normal Conditions			
Organic Tannin Gels			
Tannin + resorcinol + formaldehyde + sodium dodecyl sulfate (TRFSDS)	Preparation in oven: Hydrogels prepared at 85 °C for 72 h (gelation) and 48 h (curing); Preparation in a microwave: 85 °C for 3 h (gelation and curing) and 1–2 h (drying); NaOH and SDS were used as catalyst and surfactant, respectively; Initial pH of 5.5; SDS amount of 5 wt.%; Mass ratio of (T + R)/F of 1.2; MFTS of 25 wt.%; T was used in different proportions (0–100 wt.%) to replace R	Organic xerogels; Porosity = 20–85% The synthesis of tannin xerogels in a microwave oven allowed a reduction in time of 95%, including the time from sol-gel reaction until complete drying. Resorcinol could be replaced by tannin in the production of organic xerogels. However, the use of surfactant is always required because the structure collapses in the presence of large quantities of tannin (>25 wt.%). Materials with high proportion of tannin (75 wt.%) were produced with very small difference of porosity through conventional and microwave synthesis conditions (77% and 62%, respectively).	[124]
Carbon tannin gels			
Tannin microspheres (TM)	Microspheres prepared at 80 °C for 1 h stirred at 200–1200 rpm; Mixture of surfactant (SPAN 80) (0–10%) and sunflower oil; Initial pH of 4.3; MFTS of 40 wt.%; Carbonization: 900 °C for 2 h; Heating rate of 2 °C/min	Carbon xerogels; S_{BET} = 7–666 m^2/g Microspheres were prepared by inverse emulsion polymerization in sunflower oil, followed by subcritical drying and carbonization. Higher stirring speed produced microspheres with low average size due to the shear rate that lead particles (microspheres) at smaller sizes. At the same time, smaller microspheres were produced by increasing the surfactant content and promoting greater contact between oil and the aqueous phase. This process formed small drops of resin that became microspheres after gelation. However, there was a limit amount of surfactant and speed stirring to produce small microspheres of 5 wt.% and 500 rpm, respectively. The mean deviation decreased with the increasing speed stirring (with or without surfactant) and the final microspheres presented narrow micropore size distribution centered at 0.4–0.5 nm.	[125]
Tannin + formaldehyde (TF)	Hydrogels prepared at 85 °C for 120 h; pH from 3.3 to 8.3; Supercritical drying in acetone; Carbonization: 900 °C for 2 h; Heating rate of 5 °C/min	Carbon aerogels; S_{BET} = 580–720 m^2/g; Porosity = 75–99% Carbon aerogels based on tannin-formaldehyde solutions were prepared in a broad pH range. The microstructure of gels presented the typical nodular shape of phenolic aerogels, where their size was determined by the initial pH of the solution. The polymerization reaction was performed with methylene or methylol bridges, as showed by ^{13}C NMR analysis. The final material presented high mesopore fraction (57–78%). The synthesized TF aerogels were five times cheaper than resorcinol-formaldehyde aerogels.	[77]
Tannin + resorcinol + formaldehyde (TRF)	Hydrogels prepared at 50 °C for 10–160 min (depending on gelation time); Acetic acid or orthophosphoric acid were used to change the initial pH of the solution (8.15); pH = 2–8; Freeze-drying; Carbonization: 900 °C for 2 h; Heating rate of 5 °C/min	Carbon cryogels; S_{BET} = 30–650 m^2/g; Porosity = 32–80% Low cost cryogels were produced by replacing two third of resorcinol by tannin, which resulted in a reduction of the final cost of the gel by a factor of 2. Also, freeze-drying represents a low-cost drying method compared to supercritical drying. The S_{BET} and porosity decreased by increasing pH. Furthermore, S_{BET} values were comparable to resorcinol-formaldehyde cryogels and phenol-formaldehyde aerogels, which are gel materials based on more expensive precursors.	[53]
Tannin + formaldehyde (TF)	Hydrogels prepared at 85 °C for 120 h; Acetic acid or NaOH were used to change the initial pH of the solution (4.3); pH = 3.3–7.3; Freeze-drying; Carbonization: 900 °C for 2 h; Heating rate of 5 °C/min	Carbon cryogels; S_{BET} = 399–1420 m^2/g; Porosity = 94–96% Carbon cryogels presented high pore volumes, mostly distributed in macro- and micropores sizes. These materials also presented very low bulk densities (0.078–0.101 g/cm^3), mostly composed by macropores (0.13–0.55 cm^3/g). Materials prepared at pH higher than 6 presented higher S_{BET} (1228–1420 m^2/g) and higher micropore volumes. The electrochemical performances of those materials were tested, and the main results are presented in Table 3.	[126]

Table 1. *Cont.*

Precursors	Conditions	Main Findings	Ref.
		Hydrogels Prepared Under Normal Conditions	
		Organic Tannin Gels	
Tannin + pluronic + formaldehyde (TPF)	Hydrogels prepared at 85 °C for 120 h; pH from 2 to 10; Subcritical drying in two steps: air and 80 °C; Mass ratio of T/P of 0.5–2; Carbonization: 900 °C for 2 h; Heating rate of 5 °C/min	Carbon xerogels; S_{BET} = 5–877 m^2/g; Porosity = 46%–86% The use of pluronic allowed the synthesis of the first carbon tannin based xerogels presenting density values in a range of 0.28–0.33 g/cm^3, which are comparable to those of carbon aerogels derived from expensive precursors. Those densities are related to samples prepared at pH lower than 7 and low amounts of pluronic (Mass ratio of T/P of 2). Narrow bimodal macro-microporous or meso-microporous xerogels were produced as a function of pH and surfactant. For lower pluronic concentrations, the micropore size distribution remained centered at 0.5 nm, while the macroporosity was shifted as a function of both pluronic and pH. By increasing the amount of pluronic and keeping the same pH, the porosity changed from a macro-microporous structure to a meso-microporous structure. The final structure was highly porous with narrow non-ordered porosity.	[98]
Tannin (T) (autocondensation)	Tannin was added to solutions of Na$_2$SiO$_3$ (35 wt.%); Mass ratio of SiO$_2$/T of 0.3–1.04; NaOH (2 mol/L) or HF (40 wt. %) were used as etching agents; Curing and ageing were performed at ambient temperature for 3 h; Supercritical drying in CO$_2$; Carbonization: 900 °C for 2 h; Heating rate of 5 °C/min	Carbon aerogels; S_{BET} NaOH = 451–709 m^2/g; S_{BET} HF= 542–783 m^2/g; Porosity = 67%–82% Porous carbon gels were prepared with no formaldehyde addition. Na$_2$SiO$_3$ was used as catalyst for auto-condensation and as template for porosity formation. The final porous properties were controlled by silica/tannin ratio, while NaOH or HF were used as etching agents. XPS and Raman analyses showed the non-ordered aromatic structure of tannin-based carbon aerogels. Carbon aerogels washed with HF solutions presented better porosity (71%–82%) compared to NaOH (67%–75%). Samples prepared at low SiO$_2$/T mass ratio (0.30–0.75) had no significant differences. However, the highest SiO$_2$/T mass ratio (1.04) presented the lowest density, and consequently the highest porosity. By increasing the quantity of silica in the reactional medium, there was a decrease in the microporosity associated to S_{BET}, whereas the highest values were found for SiO$_2$/T = 0.45. Mesopores were better developed at SiO$_2$/T mass fraction of 0.75 (1.71 cm^3/g for samples etched with HF).	[127]
Tannin + sodium dodecyl sulfate + formaldehyde (TSDSF)	Hydrogels prepared at 85 °C for 72 h; NaOH and sodium dodecyl sulfate (SDS) were used as catalyst and surfactant, respectively; Initial pH of 5.5, SDS amount of 5–20 wt.% and T/F mass ratio of 0.6–2.6; MFTS of 25 wt.%; Supercritical drying in CO$_2$; Carbonization: 900 °C for 2 h; Heating rate of 5 °C/min	Carbon xerogels; Porosity = 25%–78% There was an optimal amount of surfactant (10 wt.%) for producing highly porous carbon xerogels (density of 0.34 g/cm^3 and porosity of 78%). However, when using 10 wt.% of surfactant and increasing the T/F mass ratio, the pore volumes increased, and the average pore size decreased. Also, the concentration of formaldehyde was important for attaining materials with greater porosity, since an ideal proportion of crosslinker promotes the creation of links between surrounding clusters.	[97]
Tannin + formaldehyde (TF)	Hydrogels prepared at 85 °C for 120 h; pH of 2, 5 and 8; Activation with NaOH or KOH; Mass ratio of NaOH/gel of 0.045, 0.181 and 0.724 (on a dry basis); Mass ratio of KOH/gel of 0.063, 0.253 and 1.013 (on a dry basis); Activation: 750 °C for 1 h; Heating rate of 5 °C/min	Activated carbon xerogels (ACX); S_{BET} = 50–1810 m^2/g The exchange step was performed with alkali solutions at different concentrations. Higher concentrations of alkali (NaOH or KOH) induced the development of surface area. In addition, the microstructure of ACX changed from nodular to microcellular carbon foam and, the effect of the pH was not meaningful in these different structures. A smaller dispersion of particle sizes and high surface areas related to micropores volumes were obtained using smaller mass ratio of alkalis (0.045–1.013), compared to usual values in the literature (>1). Probably, the soft structure of gels allowed the alkali solution to interpenetrate in its three-dimensional structure, causing greater alkali diffusion and favoring the formation of higher porosity.	[118]

Table 2. Organic and carbon tannin gels synthesized under hydrothermal conditions.

Precursors	Conditions	Main Findings	Ref.
		Hydrogels Prepared Under Hydrothermal Conditions *Organic and Carbon Hydrothermal Tannin Gels*	
Evaporated aminated tannin (EAT)	Hydrothermal gel: Amination at room temperature followed by evaporation and HTC of the solid material in water at 180 °C for 24 h; Subcritical drying in two steps: air and 80 °C; Carbonization: 900 °C for 3 h; Heating rate at 1 °C/min.	Xerogel: S_{BET} = 32 m^2/g This amination process under HTC conditions lead to a gel without any crosslinker agent. The mechanism was based on autocondensation and partly dehydrated tannin mainly through the heterocycle opening. Materials also had high percentage of nitrogen (up to 3.4%). Carbon xerogel: S_{BET} = 500 m^2/g The synthesized carbon gel had a porous structure comprised of a mixture of wider microporosity and mesoporosity (Isotherms Type I and IV). Materials had high percentage of nitrogen (up to 2.9%) and a micropore/total pore volume ratio of 72%. XPS analysis showed several forms of nitrogen such as pyridinic (N-6) and neutral amines in the hydrothermal xerogel, whereas pyridic (N-5) and oxidized (N-X) (stable forms of nitrogen) were found in its carbon form.	[105]
Evaporated aminated tannin (EAT)	Hydrothermal gel: Amination at room temperature followed by evaporation and HTC of the solid material in water at 180–220 °C for 24 h; Subcritical drying in two steps: air and 80 °C; Carbonization: 900 °C for 3 h; Heating rate at 1 °C/min.	Xerogel: S_{BET} = 64–102 m^2/g A monolithic gel was obtained under HTC at temperatures from 180 to 220 °C. Not many differences in relation to the physicochemical properties were observed in this range of temperature. These gels were nitrogen rich materials (up to 4.3%). Carbon xerogel: S_{BET} = 311–552 m^2/g Higher HTC temperatures lead to lower surface areas (~ 300 m^2/g) due to the formation of bigger and less porous spherical nodules. After carbonization, materials still had a high percentage of nitrogen (up to 2.9%). N-doped materials were successfully synthesized for electrochemical devices. These results were presented and discussed in Table 3.	[109]
Evaporated aminated tannin (EAT)	Hydrothermal gel: Amination at room temperature followed by evaporation and HTC of the solid material at mass fractions of 11, 18 and 27% in water at 180 °C for 24 h. The final gels were first soaked in ethanol for 3 days, and then subcritically (air and 80 °C), supercritically (CO$_2$), and freeze-dried; Carbonization: 900 °C for 3 h; Heating rate at 1 °C/min.	Xerogel, aerogel and cryogel: S_{BET} = 102–295 m^2/g For the most diluted gels (11 wt.%), the porosity improved in this sequence: xerogel, cryogel and aerogel. For the least diluted gels, similar surface areas were obtained at any kind of drying method. The mechanism of gel formation was described in Section 2.4. Carbon xerogel, aerogel and cryogel: S_{BET} = 496–860 m^2/g Unlike organic gels, carbon gels prepared with the highest mass fraction (27 wt.%) had the highest surface area, up to 900 m^2/g. It was then noticed that the most porous organic gels do not lead to carbon gels with the highest surface areas.	[111]
Tannin (T)	Hydrothermal gel: Tannin aqueous solution under HTC conditions at 180 °C for 24 h; pTSA (para-toluenosulfonic acid) was used to change the pH from 4.2 to 1; Subcritical drying in two steps: air and 80 °C; Carbonization: 900 °C for 3 h; Heating rate at 1 °C/min.	Xerogel: S_{BET} = 1.25 m^2/g Monolith material was obtained by reducing pH, resulting in a positive impact on the hydrochar yield. However, a very low surface area was obtained. Carbon xerogel: S_{BET} = 796 m^2/g Monolithic materials having spherical nodules at different diameters were obtained: 2.9 μm at pH 2; and 4.6 μm at pH 4.2. This trend was different from phenolic gels prepared with tannin-formaldehyde. The surface area of 796 m^2/g was attained for the material prepared at lowest pH. Its higher microsphere nodule diameter might have promoted the gasification of low molecular weight gases, which enhanced the development of their surface area.	[106]
Evaporated aminated (pine) tannin (EAT)	Hydrothermal gel: Amination at room temperature followed by evaporation and HTC of the solid material in water at 180 °C for 24 h; Subcritical drying in two steps: air and 80 °C; Carbonization: 900 °C for 3 h; Heating rate at 1 °C/min.	Xerogel: Compacted monoliths having homogeneous spherical particles typical of gels. Carbon xerogel: S_{BET} = 485 m^2/g Micro-mesoporous carbon gels exhibited high surface area with high nitrogen and oxygen functionalized groups, which were applied as electrodes for electrochemical devices. These results were described in Table 3.	[128]
Bayberry tannin and graphene oxide (GO)	Hydrothermal gel: A mixture of graphene oxide and tannin at different mass ratios (1:0–1:5); Sonication for 30 min and HTC at 120 °C for 12 h; Gels were soaked in HNO$_3$ solution for 4 days; Freeze-drying.	Hydrothermal cryogel: S_{BET} = 75 m^2/g The monolith had a three-dimensional structure, consisting of crosslinked GO sheets, and tannin. The gels showed a type III isotherm characteristic of mesoporous materials, having surface area up to 75 m^2/g. Its application as adsorbent to remove Sr^{2+} from water is described in Table 3	[112]

2.6. Applications of Organic and Carbon Tannin Gels

Generally, gels have been used for a great deal of applications such as diapers, contact lenses, medical electrodes, breast implants, paints, coatings, adhesives, drug delivery, controlled release of different molecules, adsorbents, and columns for materials separation (chromatography), separation technology (ion exchange), biosensors, catalysis, actuation, and sensing through artificial muscles [129]. However, only a few applications of organic and carbon gels made from tannin are presented in Table 3, according to the available literature. The removal of metals (e.g., Cr^{2+}, Cu^{2+}, Pb^{2+}, and Zn^{2+}) was investigated by the usage of tannin-gel type already explained in Section 1.3 (Tannin complexation of metals). They are less porous than gels described in this review, as they did not follow the same procedure of ageing, solvent exchange, and drying methods. Despite their lower porosity, tannin-gels have been reported to be mostly applied as adsorbents for the removal of metal contaminants from water at a concentration above the threshold fixed by environmental regulators. This is because tannin presents phenolic hydroxyl molecules that have a specific affinity to metal ions, and the ability to chelate them. An example of the tannin chelation process is presented in Figure 4. The advantage of using tannin-gel as adsorbent over tannin is that a gel type material is insoluble in water whereas tannin powder is very soluble, and it must be insolubilized or immobilized after the water treatment process.

It is also noticed that the applications of tannin gels are proposed depending on the gel type, namely aero-, cryo- and xero-gel. For example, two types of organic aerogels made from tannin-soy-formaldehyde and tannin-formaldehyde were tested as thermal insulators. This is because of their very low density (0.05–0.84 g/cm^3) and high mesopore volumes (up to 2.3 cm^3/g) [54,60]. The first aerogel presented a thermal conductivity of 0.033 W/mK, which was higher than that of air (0.026 W/mK). The authors reported that even though high porosity was attained, its volume of mesopores was only half of the total pores volume thus, narrower pores and fewer macropores would be necessary to reduce its thermal conductivity [54]. The second aerogel (tannin-formaldehyde), however, presented a thermal conductivity close to the air (0.027 W/mK) due to its low density and a significant fraction of very narrow pores. Additionally, the first aerogels showed a filamentous structure different from the typical "string-of-pearls" found for tannin-formaldehyde gel and thus, the thermal conductivity might have presented a distinct mechanism for both microstructures. This finding demonstrated that depending on the gel type and its physicochemical and textural characteristics (density, pores distribution, surface area, etc.), they can be selected for very specific applications.

For example, mesoporous materials are very interesting for water treatment because an appropriate pore size distribution is crucial in order to adsorb molecules of different sizes. Indeed, organic molecules (high-size molecule) were found to be successfully adsorbed by materials with larger micropores and mesopores [130]. Bimodal porosity was also found to be interesting for various applications because of the accessibility of ions or compounds to be transported from wider pores to micropores. On the other hand, micropores were found to play an important role in enhancing the specific capacitance. However, micropores with complex pore structure are not very useful for their capacity to store electrical energy. For example, pores lower than 0.7 nm may contribute to improving the specific capacitance, but they are not accessible at a high discharge rate [131]. Thus, the pore structure and pore size distribution must be optimized in order to have materials with high capacitance performances.

Researchers were motivated to explore the electrochemical properties of hydrothermal tannin carbon gels not only because of their great developed porosity, but also due to the enhancement of their functional groups, especially nitrogen and oxygen connected to their surface. This is because supercapacitor performances depend on the surface area accessible to the electrolyte ions, the presence of mesopores, and some heteroatoms, such as oxygen and nitrogen for the improvement of wettability and electronic conductivity [132–134]. Indeed, the final evaporated aminated tannin under HTC conditions presented nitrogen compounds in their structure, which had an important influence on the capacity of such materials to store electrical energy. According to Table 3, these materials presented the highest specific capacitance (up to 390 F/g) at scan rate of 2 mV/s [111]. Such values were higher than those reported for biosourced carbon gels in the available literature [126,135–137]. This finding is due to the combination of functional groups of oxygen and nitrogen that contributed to the enhancement of pseudo-capacitance through faradaic effects [60,111]. However, the authors highlighted that to achieve even better electrochemical performances at higher scan rates, the mesostructuration within 3 and 13 nm must be created in the structure of the carbon gel for the accessibility of micro- or ultramicro-pores. So, further studies are still needed for an effective application of tannin carbon gels in energy storage.

Table 3. Applications of organic and carbon gels made from tannin.

Application	Type of Gel	Testing Conditions	Main Findings	References
Organic tannin gels				
Thermal insulator	Aerogel	Thermal conductivity of soy-tannin-formaldehyde in a cylindrical shape was performed at room temperature	Tannin gel conductivity is higher than that of air (0.033 W/mK compared to 0.026 W/mK). The presence of narrow mesopores and fewer macropores would be required to improve their performance as thermal insulators.	[54]
Thermal insulator	Aerogel	Thermal conductivity of tannin-formaldehyde in a cylindrical shape was performed at room temperature	Tannin gel conductivity is close to that of air (0.027 W/mK). Low thermal conductivity was reported for gels with low density and very narrow pores.	[60]
Adsorbent for water treatment				
Metal Pb^{2+}	Tannin-gel	Synthetic effluent 1000 mg/L, batch tests; ratio metal: adsorbent 0.1 g: 100 mL; Equilibrium time 5 h, pH = 1–7, T = 20 °C	Tannin-gel behavior as ionic exchanger: two Na^+ ions exchanged by one Pb^{2+} ion. Pb removal efficiency increased from 58 to 115 mg Pb/g with increased pH from 3 to 4.2, respectively.	[84]
Metal Cr^{6+}	Tannin-gel	Synthetic effluent 1000 mg/L, batch tests; ratio metal: adsorbent 2.5 g:100 mL; Equilibrium time 30 min, pH = 0.85–5, T = 30 °C	Adsorption mechanism consists of four steps: 1) Esterification of chromate with tannin molecules; 2) Reduction of Cr^{6+} into Cr^{3+}; 3) Formation of carboxyl groups by tannin oxidation; and 4) Ion exchange of Cr^{3+} with carboxyl and hydroxyl groups. Maximum adsorption capacity: 287 mgCr/g at pH 2.	[83]
Metal Cr^{6+}	Tannin-gel	Synthetic effluent 500–5000 mg/L, batch tests; ratio metal: adsorbent 0.2 g:100 mL; Equilibrium time 8 h, pH = 1–12, T = 25–45 °C	Cr adsorption reached the maximum value of 488 mg/g at pH 1 (25 °C). The mechanism of Cr adsorption is based on: 1) Cr(IV) adsorption by phenolic groups through chromate esterification with tannin-gel surface; 2) Cr(VI) reduction to Cr(III); 3) Carboxylate group formation due to tannin-gel oxidation; 4) Cr(III) retention on tannin-gel surface; and finally 5) Cr adsorption through hydroxyl and carboxyl groups.	[138]
Metal Cu^{2+}	Tannin-gel	Synthetic effluent 10–150 mg/L, batch tests; ratio metal: adsorbent 0.1 g:100 mL; Equilibrium time 3 h, pH = 2–5, T = 25 °C	Adsorption mechanism is a result of ion exchange or complexation between Cu^{2+} ions and phenolic groups present on tannin-gel surface. Adsorption decreases at lower pH due to ion exchange equilibrium. Maximum adsorption capacity: 44 mgCu/g at pH 5.	[87]
Metal Zn^{2+}	Tannin-gel	Synthetic effluent, batch tests; Equilibrium time 2 weeks, pH = 7, T = 20 °C	Tannin-gel made from lab-extracted Pine tannin presented the best performance for Zn removal. Maximum adsorption capacity: 65 mgZn/g.	[88]
Metal Ni^{2+}	Tannin-gel	Synthetic and real effluent (synthetic and river water) 50–200 mg/L, batch tests; ratio metal: adsorbent 0.25–1 g: 100 mL; Equilibrium time 35 min, pH = 2–7, T = 23 °C	The adsorption of Ni ions took place in a homogeneous tannin gel surface (monolayer adsorption). Maximum adsorption capacity: 250 mgNi/g at pH 5	[139]
Metal Sr^{2+}	Hydrothermal cryogel	Synthetic effluent 10–150 mg/L, batch tests; ratio metal: adsorbent 0.02 g:100 mL; Equilibrium time 10 h, pH = 9, T = 25 °C	Graphene oxide-tannin gel prepared under HTC conditions showed excellent adsorption performance for the removal of Sr^{2+} (68 mgSr/g). Surface chemical analysis showed that Sr^{2+} was largely dependent on oxygen functional groups, pH, salinity and ionic strength.	[112]
Rare metal V	Tannin-gel	Synthetic effluent 0.2 mM, batch tests; ratio metal: adsorbent 0.02 g:100 mL; Equilibrium time 1 h, pH = 1–8, T = 30 °C	V was efficiently adsorbed from different solutions: $VOCl_2$ and NH_4VO_3. Stable compounds were formed between VO^{2+} (acid character) and catechol and pyrogallol (alkali behavior). V adsorption from NH_4VO_3 was based on adsorption of H_3VO_4 (pH = 3.75) and reduction of VO^{2+} to VO_2^- at pH 6.	[140]

Table 3. Cont.

Application	Type of Gel	Testing Conditions	Main Findings	References
Rare metal Au	Tannin-gel	Synthetic effluent 10 mg/L, batch tests; ratio metal: adsorbent 0.01 g:100 mL; pH = 2–3.8, T = 20 °C	Adsorption of Au took place through the reduction of trivalent Au ions into metallic Au as well as oxidation of hydroxyl groups present in tannin-gel to carbonyl groups. Maximum adsorption capacity: 8000 mgAu/g	[85]
Rare metal Au	Tannin-gel	Synthetic effluent 0.1 mM, column tests; Flow bed at 150 and 300 mL/h; pH = 2–6	98.5% of Au from HAuCl$_4$ was adsorbed at pH 6. The mechanism of Au adsorption was based on: 1) Ligand exchange: AuCl$_4^-$ with hydroxylphenyl groups present in the tannin-gel; 2) Reduction of Au(III) to Au(0); and 3) Adsorption of Au(0).	[141]
Rare metal Pd(II)	Tannin-gel	Synthetic effluent 10 mg/L, batch tests; ratio metal: adsorbent 0.04 g:100 mL; pH = 1.3–2.5, T = 25 °C	Adsorption of Pd(II) was based on the inner sphere redox reaction: Pd(II) ions were adsorbed as metallic Pd; hydroxyl groups were oxidized; and a ligand-substitute Pd(II) tannin inner sphere complex was formed.	[86]
Rare metal Pd(II)	Aminated tannin-gel freeze-dried	Synthetic effluent 100 mg/L, batch tests; ratio metal: adsorbent 0.1 g:100 mL; Acidic medium, T = 25 °C	Adsorption of Pd(II) was due to metal ions complexation and/or electrostatic interaction. Also, acidic metal ions had high affinity towards amine basic groups. Maximum adsorption capacity: 80 mgPd/g.	[142]
Rare metals Pd and Pt	Aminated tannin-gel freeze-dried	Synthetic effluent 0.001 M, batch tests, single and multiple metal solutions; ratio metal: adsorbent 0.1 g:100 mL; Equilibrium time 20 h, pH = 0–5, T = 25 °C	The adsorption of Pd and Pt on tannin-gel surface increased with increasing pH and temperature, and with decreasing chloride ion concentration. The amino groups presented in tannin-gel formed stable complex with metal ions but the adsorbability of Pd(II) was much higher than Pt(IV). Interesting that aminated tannin gel adsorbed mostly Pd(II) from mixed solutions even though it had good adsorbability for Pt(IV) from single metal solution	[143]
Rare metals Au, Pd, Pt and Rh	Tannin-gel	Synthetic effluent 1 mmol/L, batch tests, single and multiple solutions; ratio metal: adsorbent 0.1 g:100 mL; pH = 1, T = 25 °C	The predominant species of each metal were adsorbed by controlling the pH (equal to 1) as well as the redox potential differences between metals and tannin-gel. Au was selectively adsorbed and reduced because its redox potential was higher than that of tannin-gel. However, the other precious metals had much lower redox potential than that of tannin-gel.	[144]
Metals: Au(III), Pd(II), Pt(IV), Cu(II), Fe(III), Ni(II), Zn(II)	Tannin-gel	Synthetic and real effluent, batch and column tests, single and multiple metal solutions; ratio metal: adsorbent 0.1–2 g:100 mL; Flow bed at 5 mL/h; Equilibrium time 12 h, acidic pH, T = 30 °C	Rare metals were efficiently adsorbed through both batch and column tests. Tannin-gel selectively adsorbed Au(III), Pd(II) and Pt(IV) over other metals: Cu(II), Fe(III), Ni(II) and Zn(II). The mechanism of adsorption of precious metals was the combination of ion exchange, electrostatic interaction and coordination with thiocarbonyl group. Au(III) was reduced to elemental Au through abundant polyphenolic groups on tannin molecule. Tannin-gel was regenerated under acid solution up to five times.	[145]
Boron (B)	Aminated tannin-gel freeze-dried	Synthetic effluent 200 mg/L, batch tests; ratio contaminant: adsorbent 0.5 g:100 mL; Equilibrium time 20 h, pH = 8.8, T = 30 °C	Aminated and non-aminated gels efficiently adsorbed B at pH > 7. The adsorption of B took place through the chelate formation of tetrahydroxyborate ion and the hydroxyl and amino groups presented in tannin-gels. The adsorption capacity of the aminated cryogel was higher than that of non-aminated one due to the stable bonds between boron and nitrogen from amino groups.	[146]
Phosphate (P)	Tannin-gel freeze-dried	Synthetic effluent 100 mg/L, batch tests; ratio contaminant: adsorbent 0.5 g:100 mL; pH = 2–12, T = 25 °C	The gel impregnated with iron and oxidized with nitric acid showed adsorption selectivity for phosphate. The adsorption process was independent of the pH (from 3 to 12). Maximum adsorption capacity: 31 mgP/gFe	[147]

Table 3. Cont.

Application	Type of Gel	Testing Conditions	Main Findings	References
Organic MB	Tannin-gel	Synthetic effluent, batch tests; Equilibrium time 2 weeks, pH = 7, T = 20 °C	Tannin-gel made from lab-extracted Pine tannin presented the best performance for MB removal. Maximum adsorption capacity: 432 mgMB/g	[88]
Organic MB	Tannin-gel	Synthetic effluent 1000 mg/L, batch tests; Equilibrium time 15 days; pH = 4–10, T = 20 °C	Adsorption of MB was improved by increasing pH, probably because the dye appeared with a higher cationic degree and thus, it enhanced electrostatic interactions. Maximum adsorption capacity: 483 mgMB/g	[148]
Organic BR	Tannin-gel	Synthetic effluent 40 mg/L, batch tests; ratio contaminant: adsorbent 0.04 g:100 mL; Equilibrium time 1.5 h, pH = 2–8, T = 28 °C	Good adsorption of BR due to the presence of functional groups on tannin-gel structure: phenolic, carboxylic, alcoholic, ether and aromatic rings. Maximum adsorption capacity: 45 mgBR/g	[149]
Organic CTAB	Tannin-gel	Synthetic effluent, batch tests; Equilibrium time 2 weeks, pH = 7, T = 20 °C	Tannin-gel made from lab-extracted Pine tannin presented the best performance for CTAB removal. Maximum adsorption capacity: 773 mgCTAB/g	[88]
Benzene and toluene	Tannin-gel	Synthetic effluents 1% sol, batch tests; Equilibrium time 1 h; ratio contaminant: adsorbent 0.1 g:100 mL; pH = 2–8.6, T = 60 °C	The removal of toluene was more effective than benzene probably because of the interactions between the methyl groups on toluene and the OH groups on tannin gel. Results show up to 99% removal of toluene and benzene after 30 min batch tests.	[150]

Carbon tannin gels

Application	Type of Gel	Testing Conditions	Main Findings	References
Thermal insulator	Carbon xerogel	Thermal conductivity of tannin-formaldehyde-surfactant in a cylindrical shape was performed at room temperature	Carbon tannin gel conductivity is higher than that of air (0.039 W/mK compared to 0.026 W/mK). The presence of narrow mesopores would be required to improve its performance as thermal insulator.	[97]

Electrochemistry

Application	Type of Gel	Testing Conditions	Main Findings	References
Supercapacitor	Carbon cryogel	Supercapacitor device based on a three-electrode cell configuration with an aqueous acid electrolyte (4 M H_2SO_4)	Carbon cryogels prepared at pH higher than 6 had low density and high surface areas (up to 1200 m^2/g). Thus, such materials as electrodes for supercapacitor reached capacitances as high as 100 F/g (scan rate 2 mV/s). In addition to mesopores, ultra- and supermicropores played an important role on their performance as electrodes for supercapacitor.	[126]
Supercapacitor	Hydrothermal carbon aero-, cryo- and xerogel	Supercapacitor device based on a three-electrode cell configuration with an aqueous acid electrolyte (4 M H_2SO_4)	Carbon aerogel, cryogel, and xerogel prepared from HTC of evaporated aminated tannin (MFTS of 27 wt.%) reached surfaces areas of 860, 754, and 585 m^2/g and specific capacitances of 362, 387, and 330 F/g (scan rate 2 mV/s), respectively. The presence of nitrogen (2–3 wt.%) and oxygen (17–18 wt.%) functional groups played an important role on their performance for electrical energy storage, especially through pseudo-capacitance. However, mesostructuration within 3–13 nm should be created to improve the capacitance reduction at a higher scan rate.	[111]
Supercapacitor	Hydrothermal carbon xerogel	Supercapacitor device based on a three-electrode cell configuration with an aqueous acid electrolyte (1 M H_2SO_4)	Hydrothermal carbon xerogel made from evaporated aminated Pine tannin reached a surface area and a specific capacitance of 485 m^2/g and 253 F/g (scan rate 0.5 mV/s), respectively. The material presented high concentration of nitrogen and oxygen functional groups (6 mmol/g) that played an important role on their performance as electrodes for a supercapacitor.	[128]

MB: Methylene blue; BR: Brilliant red; CTAB: Cetyltrimethylammonium bromide.

3. Conclusions

The present study reviewed the pertinent literature dedicated to the synthesis, physicochemical and textural properties, and applications of organic and carbon tannin gels. The main findings are outlined below:

- Tannins are the most abundant compounds from biomass after cellulose, hemicellulose, and lignin that can be used in a sustainable biorefinery plant. Among them, condensed polyflavonoid tannins exhibit a complex and heterogeneous structure with highly reactive sites in their structure that lead to condensation, polymerization or rearrangement under acid or alkaline conditions, in the presence of catalysts, metals, sulfite, and especially aldehydes (e.g., formaldehyde), which are efficient precursors for the preparation of solid gels.
- The pH of the solution may also affect the final gels obtained under normal or hydrothermal conditions. Under normal conditions, gels prepared at low pH present high nodules, consequently high clusters, and high porosity, whereas at high pH, gels presented smaller nodules, reduced porosity, and narrow pores. In the case of hydrothermal gels, at low pH, gels attained low spherical nodules with higher porosity compared to bigger spherical particles of lower porosity at high pH. In the last case, smaller nodules favored the evolution of volatile matter during carbonization, and consequently improved the surface area and porosity development.
- The mass fraction of total solids from initial solutions has also an impact on final properties of tannin gels. Basically, very diluted systems lead to more porous gels, whereas high mass fractions tend to produce less porous materials. However, the most porous gels bear more capillary stresses during the drying step, which can be controlled by the addition of a surfactant in the medium.
- During tannin gel synthesis, the drying method defines the porous structure of the organic tannin gel, and consequently its application. Although aero- and cryo-gels are produced from costly techniques, their developed porous structure due to lower shrinkage is desired for several applications, which have not been tested or fully explored as yet (e.g., catalysis, column separation methods, chromatography, and thermal insulation).
- Carbonization and activation conditions play an important role in the porosity development of tannin gels. Optimal conditions of temperature, residence times, heating rate, and physical and chemical agents are suitable for the development of their porosity for specific applications. Thus, thermal treatment parameters conditions must be optimized during the production of tannin carbon gels.
- Tannin gels present interesting physicochemical properties for several applications. However, further studies should be performed to investigate their pore size distribution and the development of mesopores, which are crucial for most of the applications discussed: the removal of contaminants from wastewater, thermal insulation and energy storage. In addition, most studies on the application of tannin gels were based on small-scale tests.
- Thus, tannin organic and carbon gels are based on non-toxic, biosourced, and low-cost materials, which have textural properties (porosity or surface area) that are comparable to synthetic phenolic precursors such as phenol and resorcinol-formaldehyde gels.

Funding: This research received no external funding.

Conflicts of Interest: The authors declare no conflict of interest.

References

1. Arbenz, A.; Avérous, L. Chemical modification of tannins to elaborate aromatic biobased macromolecular architectures. *Green Chem.* **2015**, *17*, 2626–2646. [CrossRef]
2. Haslam, E. *Plant. Polyphenols: Vegetable Tannins Revisited*; Chemistry and Pharmacology of Natural Products; Cambridge University Press: Cambridge, UK, 1989.

3. De Hoyos-Martínez, P.L.; Merle, J.; Labidi, J.; Charrier-El Bouhtoury, F. Tannins extraction: A key point for their valorization and cleaner production. *J. Clean. Prod.* **2019**, *206*, 1138–1155. [CrossRef]
4. Pizzi, A. Tannin-based adhesives. *J. Macromol. Sci. Polymer Rev.* **1980**, *18*, 247–315. [CrossRef]
5. Pizzi, A. Recent developments in eco-efficient bio-based adhesives for wood bonding: Opportunities and issues. *J. Adhes. Sci. Technol.* **2006**, *20*, 829–846. [CrossRef]
6. Roux, D.G.; Paulus, E. Condensed tannins. 8. The isolation and distribution of interrelated heartwood components of *Schinopsis* spp. *Biochem. J.* **1961**, *78*, 785–789. [CrossRef]
7. Drewes, S.; Roux, D. Condensed tannins. 15. Interrelationships of flavonoid components in wattle-bark extract. *Biochem. J.* **1963**, *87*, 167–172. [CrossRef]
8. Abdalla, S.; Pizzi, A.; Ayed, N.; Charrier, F.; Bahabri, F.; Ganash, A. MALDI-TOF and ^{13}C NMR analysis of Tunisian *Zizyphus jujuba* root bark tannins. *Ind. Crops Prod.* **2014**, *59*, 277–281. [CrossRef]
9. Drovou, S.; Pizzi, A.; Lacoste, C.; Zhang, J.; Abdulla, S.; El-Marzouki, F.M. Flavonoid tannins linked to long carbohydrate chains—MALDI-TOF analysis of the tannin extract of the African locust bean shells. *Ind. Crops Prod.* **2015**, *67*, 25–32. [CrossRef]
10. Roux, D.G.; Ferreira, D.; Hundt, H.K.L.; Malan, E. Structure, stereochemistry, and reactivity of natural condensed tannins as basis for their extended industrial application. *Appl. Polym. Symp.* **1975**, *28*, 335–353.
11. Roux, D.G. Recent advances in the chemistry and chemical utilization of the natural condensed tannins. *Phytochemistry* **1972**, *11*, 1219–1230. [CrossRef]
12. Roux, D.G. *Modern Applications of Mimosa Extract*; Leather Industries Research Institute: Grahamstown, South Africa, 1965.
13. Pizzi, A. *Advanced Wood Adhesives Technology*; Marcel Dekker: New York, NY, USA, 1994.
14. Pizzi, A. Natural Phenolic Adhesives 1: Tannin. In *Handbook of Adhesive Technology*; Pizzi, A., Mittal, K.L., Eds.; Marcel Dekker: New York, NY, USA, 2003; pp. 573–598.
15. Christiansen, A.W.; Gillespie, R.H. *Wood Adhesives in 1985: Status and Needs*; Forest Products Laboratory, Ed.; Proceedings. Forest Products Research Society; FPRS: Madison, WI, USA, 1986.
16. Pizzi, A.; Stephanou, A. Comparative and differential behaviour of pine vs. pecan nut tannin adhesives for particleboard. *Holzforsch. Holzverwert.* **1993**, *45*, 30–33.
17. McGraw, G.W.; Rials, T.G.; Steynberg, J.P.; Hemingway, R.W. Chemistry of Pecan Tannins and Analysis of Cure of Pecan Tannin-Based Cold-Setting Adhesives with a DMA 'Micro-Beam' Test. In *Plant Polyphenols. Basic Life Sciences*; Hemingway, R.W., Laks, P.E., Eds.; Springer: Boston, MA, USA, 1992; Volume 59, pp. 979–990.
18. Sealy-Fisher, V.J.; Pizzi, A. Increased pine tannins extraction and wood adhesives development by phlobaphenes minimization. *Holz. Als. Roh. -Werkst.* **1992**, *50*, 212–220. [CrossRef]
19. Roux, D.G. *Wattle Tannin and Mimosa Extract*; Leather Industries Research Institute: Grahamstown, South Africa, 1965.
20. Pizzi, A. Sulfited tannins for exterior wood adhesives. *Colloid Polym. Sci.* **1979**, *257*, 37–40. [CrossRef]
21. Ohara, S.; Hemingway, R.W. Condensed tannins: The formation of a diarylpropanol-catechinic acid dimer from base-catalyzed reactions of (+)-catechin. *J. Wood Chem. Technol.* **1991**, *11*, 195–208. [CrossRef]
22. Pizzi, A. Pine tannin adhesives for particleboard. *Holz Als Roh- Werkst.* **1982**, *40*, 293–301. [CrossRef]
23. Pizzi, A.; von Leyser, E.P.; Valenzuela, J.; Clark, J.G. The chemistry and development of pine tannin adhesives for exterior particleboard. *Holzforschung* **1993**, *47*, 168–174. [CrossRef]
24. Valenzuela, J.; von Leyser, E.; Pizzi, A.; Westermeyer, C.; Gorrini, B. Industrial production of pine tannin-bonded particleboard and MDF. *Eur. J. Wood Wood Prod.* **2012**, *70*, 735–740. [CrossRef]
25. Pizzi, A.; Valenezuela, J.; Westermeyer, C. Low formaldehyde emission, fast pressing, pine and pecan tannin adhesives for exterior particleboard. *Holz Als Roh- Werkst.* **1994**, *52*, 311–315. [CrossRef]
26. Pizzi, A.; Stephanou, A. Fast vs. slow-reacting non-modified tannin extracts for exterior particleboard adhesives. *Holz. Als. Roh. -Werkst.* **1994**, *52*, 218–222. [CrossRef]
27. Meikleham, N.; Pizzi, A.; Stephanou, A. Induced accelerated autocondensation of polyflavonoid tannins for phenolic polycondensates. I. ^{13}C-NMR, ^{29}Si-NMR, X-ray, and polarimetry studies and mechanism. *J. Appl. Polym. Sci.* **1994**, *54*, 1827–1845. [CrossRef]
28. Slabbert, N. Complexation of condensed tannins with metal ions. In *Plant Polyphenols: Biogenesis, Chemical Properties, and Significance*; Hemingway, R.W., Laks, P.E., Eds.; Plenum Press: New York, NY, USA, 1992; pp. 421–436.

29. Tondi, G.; Oo, C.W.; Pizzi, A.; Trosa, A.; Thevenon, M.F. Metal adsorption of tannin based rigid foams. *Ind. Crops Prod.* **2009**, *29*, 336–340. [CrossRef]
30. Oo, C.W.; Kassim, M.J.; Pizzi, A. Characterization and performance of *Rhizophora apiculata* mangrove polyflavonoid tannins in the adsorption of copper (II) and lead (II). *Ind. Crops Prod.* **2009**, *30*, 152–161. [CrossRef]
31. Pizzi, A. Tannins: Prospectives and actual industrial applications. *Biomolecules* **2019**, *9*, 344. [CrossRef] [PubMed]
32. Pizzi, A.; Cameron, F.A.; Eaton, N.J. The tridimensional structure of polyflavonoid tannins by conformational analysis. *J. Macromol. Sci. Part. - Chem.* **1985**, *22*, 515–540. [CrossRef]
33. Thébault, M.; Pizzi, A.; Essawy, H.A.; Barhoum, A.; Van Assche, G. Isocyanate free condensed tannin-based polyurethanes. *Eur. Polym. J.* **2015**, *67*, 513–526. [CrossRef]
34. Pizzi, A. Tannin-based polyurethane adhesives. *J. Appl. Polym. Sci.* **1979**, *23*, 1889–1891. [CrossRef]
35. Pizzi, A. Tannin-formaldehyde exterior wood adhesives through flavonoid B-ring cross linking. *J. Appl. Polym. Sci.* **1978**, *22*, 2397–2399. [CrossRef]
36. Pizzi, A.; Stephanou, A. A ^{13}C NMR study of polyflavonoid tannin adhesive intermediates. II. Colloidal state reactions. *J. Appl. Polym. Sci.* **1994**, *51*, 2125–2130. [CrossRef]
37. Shirmohammadli, Y.; Efhamisisi, D.; Pizzi, A. Tannins as a sustainable raw material for green chemistry: A review. *Ind. Crops Prod.* **2018**, *126*, 316–332. [CrossRef]
38. Saayman, H.; Roux, D. The origins of tannins and flavonoids in black-wattle barks and heartwoods, and their associated 'non-tannin' components. *Biochem. J.* **1965**, *97*, 794–801. [CrossRef]
39. Li, X.; Basso, M.C.; Braghiroli, F.L.; Fierro, V.; Pizzi, A.; Celzard, A. Tailoring the structure of cellular vitreous carbon foams. *Carbon* **2012**, *50*, 2026–2036. [CrossRef]
40. Lacoste, C.; Basso, M.C.; Pizzi, A.; Laborie, M.-P.; Celzard, A.; Fierro, V. Pine tannin-based rigid foams: Mechanical and thermal properties. *Ind. Crops Prod.* **2013**, *43*, 245–250. [CrossRef]
41. Lacoste, C.; Basso, M.-C.; Pizzi, A.; Celzard, A.; Ella Ebang, E.; Gallon, N.; Charrier, B. Pine (*P. pinaster*) and quebracho (*S. lorentzii*) tannin-based foams as green acoustic absorbers. *Ind. Crops Prod.* **2015**, *67*, 70–73. [CrossRef]
42. Mitsunaga, T.; Doi, T.; Kondo, Y.; Abe, I. Color development of proanthocyanidins in vanillin-hydrochloric acid reaction. *J. Wood Sci.* **1998**, *44*, 125–130. [CrossRef]
43. Clark-Lewis, J.W.; Roux, D.G. Natural occurrence of enantiomorphous leucoanthocyanidian: (+)-mollisacacidin (gleditsin) and quebracho(−)-leucofisetinidin. *J. Chem. Soc.* **1959**, 1402–1406. [CrossRef]
44. Navarrete, P.; Pizzi, A.; Pasch, H.; Rode, K.; Delmotte, L. MALDI-TOF and ^{13}C NMR characterization of maritime pine industrial tannin extract. *Ind. Crops Prod.* **2010**, *32*, 105–110. [CrossRef]
45. Abdalla, S.; Pizzi, A.; Ayed, N.; Charrier-El Bouthoury, F.; Charrier, B.; Bahabri, F.; Ganash, A. MALDI-TOF Analysis of Aleppo Pine (*Pinus halepensis*) bark tannin. *BioResources* **2014**, *9*, 3396–3406. [CrossRef]
46. Ucar, M.B.; Ucar, G.; Pizzi, A.; Gonultas, O. Characterization of *Pinus brutia* bark tannin by MALDI-TOF MS and ^{13}C NMR. *Ind. Crops Prod.* **2013**, *49*, 697–704. [CrossRef]
47. Saad, H.; Charrier-El Bouhtoury, F.; Pizzi, A.; Rode, K.; Charrier, B.; Ayed, N. Characterization of pomegranate peels tannin extractives. *Ind. Crops Prod.* **2012**, *40*, 239–246. [CrossRef]
48. Hundt, H.K.L.; Roux, D.G. Condensed tannins: Determination of the point of linkage in 'terminal'(+)-catechin units and degradative bromination of 4-flavanylflavan-3,4-diols. *J. Chem. Soc. Chem. Commun.* **1978**. [CrossRef]
49. Botha, J.J.; Ferreira, D.; Roux, D.G. Condensed tannins. Circular dichroism method of assessing the absolute configuration at C-4 of 4-arylflavan-3-ols, and stereochemistry of their formation from flavan-3,4-diols. *J. Chem. Soc. Chem. Commun.* **1978**. [CrossRef]
50. Basso, M.C.; Lacoste, C.; Pizzi, A.; Fredon, E.; Delmotte, L. MALDI-TOF and ^{13}C NMR analysis of flexible films and lacquers derived from tannin. *Ind. Crops Prod.* **2014**, *61*, 352–360. [CrossRef]
51. Delgado-Sánchez, C.; Amaral-Labat, G.; Grishechko, L.I.; Sánchez–Sánchez, A.; Fierro, V.; Pizzi, A.; Celzard, A. Fire-resistant tannin–ethylene glycol gels working as rubber springs with tuneable elastic properties. *J. Mater. Chem. A* **2017**, *5*, 14720–14732. [CrossRef]
52. Brinker, C.J.; Scherer, G.W. *Sol.—gel science. The physics and chemistry of Sol.—gel processing*; Academic Press Inc.: San Diego, CA, USA, 1990.

53. Szczurek, A.; Amaral-Labat, G.; Fierro, V.; Pizzi, A.; Celzard, A. The use of tannin to prepare carbon gels. Part II. Carbon cryogels. *Carbon* **2011**, *49*, 2785–2794. [CrossRef]
54. Amaral-Labat, G.; Grishechko, L.; Szczurek, A.; Fierro, V.; Pizzi, A.; Kuznetsov, B.; Celzard, A. Highly mesoporous organic aerogels derived from soy and tannin. *Green Chem.* **2012**, *14*, 3099. [CrossRef]
55. Hench, L.L.; West, J.K. The sol-gel process. *Chem. Rev.* **1990**, *90*, 33–72. [CrossRef]
56. Pekala, R.W.; Schaefer, D.W. Structure of organic aerogels. 1. Morphology and scaling. *Macromolecules* **1993**, *26*, 5487–5493. [CrossRef]
57. Job, N.; Pirard, R.; Marien, J.; Pirard, J.-P. Porous carbon xerogels with texture tailored by pH control during sol–gel process. *Carbon* **2004**, *42*, 619–628. [CrossRef]
58. Wang, J.; Glora, M.; Petricevic, R.; Saliger, R.; Proebstle, H.; Fricke, J. Carbon cloth reinforced carbon aerogel films derived from resorcinol formaldehyde. *J. Porous Mater.* **2001**, *8*, 159–165. [CrossRef]
59. Bock, V.; Emmerling, A.; Saliger, R.; Fricke, J. Structural investigation of resorcinol formaldehyde and carbon aerogels using SAXS and BET. *J. Porous Mater.* **1997**, *4*, 287–294. [CrossRef]
60. Amaral-Labat, G.; Szczurek, A.; Fierro, V.; Pizzi, A.; Celzard, A. Systematic studies of tannin–formaldehyde aerogels: Preparation and properties. *Sci. Technol. Adv. Mater.* **2013**, *14*, 015001. [CrossRef] [PubMed]
61. Amaral-Labat, G.; Grishechko, L.I.; Fierro, V.; Kuznetsov, B.N.; Pizzi, A.; Celzard, A. Tannin-based xerogels with distinctive porous structures. *Biomass Bioenergy* **2013**, *56*, 437–445. [CrossRef]
62. Job, N.; Panariello, F.; Marien, J.; Crine, M.; Pirard, J.-P.; Léonard, A. Synthesis optimization of organic xerogels produced from convective air-drying of resorcinol–formaldehyde gels. *J. Non-Cryst. Solids* **2006**, *352*, 24–34. [CrossRef]
63. Job, N.; Théry, A.; Pirard, R.; Marien, J.; Kocon, L.; Rouzaud, J.-N.; Béguin, F.; Pirard, J.-P. Carbon aerogels, cryogels and xerogels: Influence of the drying method on the textural properties of porous carbon materials. *Carbon* **2005**, *43*, 2481–2494. [CrossRef]
64. Castro, C.D.; Quintana, G.C. Mixture design approach on the physical properties of lignin-resorcinol-formaldehyde xerogels. *Int. J. Polym. Sci.* **2015**, *2015*, 1–11. [CrossRef]
65. Aegerter, M.A.; Leventis, N.; Koebel, M.M. *Aerogels Handbook. Advances in Sol-Gel Derived Materials and Technologies*; Springer: New York, NY, USA, 2011.
66. García-González, C.A.; Alnaief, M.; Smirnova, I. Polysaccharide-based aerogels—Promising biodegradable carriers for drug delivery systems. *Carbohydr. Polym.* **2011**, *86*, 1425–1438. [CrossRef]
67. Robb, S.A.; Lee, B.H.; McLemore, R.; Vernon, B.L. Simultaneously physically and chemically gelling polymer system utilizing a poly(NIPAAm-co-cysteamine)-based copolymer. *Biomacromolecules* **2007**, *8*, 2294–2300. [CrossRef]
68. Pizzi, A.; Scharfetter, H.O. The chemistry and development of tannin-based adhesives for exterior plywood. *J. Appl. Polym. Sci.* **1978**, *22*, 1745–1761. [CrossRef]
69. Ping, L.; Brosse, N.; Chrusciel, L.; Navarrete, P.; Pizzi, A. Extraction of condensed tannins from grape pomace for use as wood adhesives. *Ind. Crops Prod.* **2011**, *33*, 253–257. [CrossRef]
70. Pizzi, A. *Wood Adhesive Chemistry and Technology*; Marcel Dekker: New York, NY, USA, 1983.
71. Fraser, D.A.; Hall, R.W.; Raum, A.L.J. Preparation of 'high-ortho' novolak resins I. Metal ion catalysis and orientation effect. *J. Appl. Chem.* **2007**, *7*, 676–689. [CrossRef]
72. Fraser, D.A.; Hall, R.W.; Jenkins, P.A.; Raum, A.L.J. Preparation of 'high-ortho' novolak resins. II. The course of the reaction. *J. Appl. Chem.* **2007**, *7*, 689–700. [CrossRef]
73. Pizzi, A. Phenolic resins by reactions of coordinated metal ligands. *J. Polym. Sci. Polym. Lett. Ed.* **1979**, *17*, 489–492. [CrossRef]
74. Pizzi, A. Phenolic and tannin-based adhesive resins by reactions of coordinated metal ligands. I. Phenolic chelates. *J. Appl. Polym. Sci.* **1979**, *24*, 1247–1255. [CrossRef]
75. Pizzi, A. Phenolic and tannin-based adhesive resins by reactions of coordinated metal ligands. II. Tannin adhesive preparation, characteristics, and application. *J. Appl. Polym. Sci.* **1979**, *24*, 1257–1268. [CrossRef]
76. Hillis, W.E.; Urbach, G. The reaction of (+)-catechin with formaldehyde. *J. Appl. Chem.* **2007**, *9*, 474–482. [CrossRef]
77. Szczurek, A.; Amaral-Labat, G.; Fierro, V.; Pizzi, A.; Masson, E.; Celzard, A. The use of tannin to prepare carbon gels. Part I: Carbon aerogels. *Carbon* **2011**, *49*, 2773–2784. [CrossRef]
78. Grishechko, L.I.; Amaral-Labat, G.; Szczurek, A.; Fierro, V.; Kuznetsov, B.N.; Pizzi, A.; Celzard, A. New tannin–lignin aerogels. *Ind. Crops Prod.* **2013**, *41*, 347–355. [CrossRef]

79. Amaral-Labat, G.A.; Pizzi, A.; Gonçalves, A.R.; Celzard, A.; Rigolet, S.; Rocha, G.J.M. Environment-friendly soy flour-based resins without formaldehyde. *J. Appl. Polym. Sci.* **2008**, *108*, 624–632. [CrossRef]
80. Lacoste, C.; Basso, M.C.; Pizzi, A.; Celzard, A.; Laborie, M.-P. Natural albumin/tannin cellular foams. *Ind. Crops Prod.* **2015**, *73*, 41–48. [CrossRef]
81. Yoshizawa, N.; Hatori, H.; Soneda, Y.; Hanzawa, Y.; Kaneko, K.; Dresselhaus, M.S. Structure and electrochemical properties of carbon aerogels polymerized in the presence of Cu^{2+}. *J. Non-Cryst. Solids* **2003**, *330*, 99–105. [CrossRef]
82. Amaral-Labat, G.; Szczurek, A.; Fierro, V.; Pizzi, A.; Masson, E.; Celzard, A. "Blue glue": A new precursor of carbon aerogels. *Microporous Mesoporous Mater.* **2012**, *158*, 272–280. [CrossRef]
83. Nakano, Y.; Takeshita, K.; Tsutsumi, T. Adsorption mechanism of hexavalent chromium by redox within condensed-tannin gel. *Water Res.* **2001**, *35*, 496–500. [CrossRef]
84. Zhan, X.-M.; Zhao, X. Mechanism of lead adsorption from aqueous solutions using an adsorbent synthesized from natural condensed tannin. *Water Res.* **2003**, *37*, 3905–3912. [CrossRef]
85. Ogata, T.; Nakano, Y. Mechanisms of gold recovery from aqueous solutions using a novel tannin gel adsorbent synthesized from natural condensed tannin. *Water Res.* **2005**, *39*, 4281–4286. [CrossRef] [PubMed]
86. Kim, Y.-H.; Ogata, T.; Nakano, Y. Kinetic analysis of palladium(II) adsorption process on condensed-tannin gel based on redox reaction models. *Water Res.* **2007**, *41*, 3043–3050. [CrossRef]
87. Şengil, İ.A.; Özacar, M. Biosorption of Cu(II) from aqueous solutions by mimosa tannin gel. *J. Hazard. Mater.* **2008**, *157*, 277–285. [CrossRef]
88. Sánchez-Martín, J.; Beltrán-Heredia, J.; Gibello-Pérez, P. Adsorbent biopolymers from tannin extracts for water treatment. *Chem. Eng. J.* **2011**, *168*, 1241–1247. [CrossRef]
89. Al-Muhtaseb, S.A.; Ritter, J.A. Preparation and properties of resorcinol-formaldehyde organic and carbon gels. *Adv. Mater.* **2003**, *15*, 101–114. [CrossRef]
90. García-González, C.A.; Camino-Rey, M.C.; Alnaief, M.; Zetzl, C.; Smirnova, I. Supercritical drying of aerogels using CO_2: Effect of extraction time on the end material textural properties. *J. Supercrit. Fluids* **2012**, *66*, 297–306. [CrossRef]
91. Fricke, J.; Tillotson, T. Aerogels: Production, characterization, and applications. *Thin Solid Films* **1997**, *297*, 212–223. [CrossRef]
92. Amaral-Labat, G. Gels Poreux Biosourcés: Production, Caractérisation et Applications. Doctoral Dissertation, University of Lorraine, Épinal, France, 2013.
93. Scherer, G.W.; Smith, D.M. Cavitation during drying of a gel. *J. Non-Cryst. Solids* **1995**, *189*, 197–211. [CrossRef]
94. Daraoui, N.; Dufour, P.; Hammouri, H.; Hottot, A. Model predictive control during the primary drying stage of lyophilisation. *Control. Eng. Pract.* **2010**, *18*, 483–494. [CrossRef]
95. Tamon, H.; Ishizaka, H.; Yamamoto, T.; Suzuki, T. Preparation of mesoporous carbon by freeze drying. *Carbon* **1999**, *37*, 2049–2055. [CrossRef]
96. Baetens, R.; Jelle, B.P.; Gustavsen, A. Aerogel insulation for building applications: A state-of-the-art review. *Energy Build.* **2011**, *43*, 761–769. [CrossRef]
97. Rey-Raap, N.; Szczurek, A.; Fierro, V.; Celzard, A.; Menéndez, J.A.; Arenillas, A. Advances in tailoring the porosity of tannin-based carbon xerogels. *Ind. Crops Prod.* **2016**, *82*, 100–106. [CrossRef]
98. Amaral-Labat, G.; Szczurek, A.; Fierro, V.; Celzard, A. Unique bimodal carbon xerogels from soft templating of tannin. *Mater. Chem. Phys.* **2015**, *149–150*, 193–201. [CrossRef]
99. Braghiroli, F.L.; Fierro, V.; Parmentier, J.; Pasc, A.; Celzard, A. Easy and eco-friendly synthesis of ordered mesoporous carbons by self-assembly of tannin with a block copolymer. *Green Chem.* **2016**, *18*, 3265–3271. [CrossRef]
100. Libra, J.A.; Ro, K.S.; Kammann, C.; Funke, A.; Berge, N.D.; Neubauer, Y.; Titirici, M.-M.; Führer, C.; Bens, O.; Kern, J.; et al. Hydrothermal carbonization of biomass residuals: A comparative review of the chemistry, processes and applications of wet and dry pyrolysis. *Biofuels* **2011**, *2*, 71–106. [CrossRef]
101. Brun, N.; García-González, C.A.; Smirnova, I.; Titirici, M.M. Hydrothermal synthesis of highly porous carbon monoliths from carbohydrates and phloroglucinol. *RSC Adv.* **2013**, *3*, 17088–17096. [CrossRef]
102. Fellinger, T.-P.; White, R.J.; Titirici, M.-M.; Antonietti, M. Borax-mediated formation of carbon aerogels from glucose. *Adv. Funct. Mater.* **2012**, *22*, 3254–3260. [CrossRef]

103. Wohlgemuth, S.-A.; White, R.J.; Willinger, M.-G.; Titirici, M.-M.; Antonietti, M. A one-pot hydrothermal synthesis of sulfur and nitrogen doped carbon aerogels with enhanced electrocatalytic activity in the oxygen reduction reaction. *Green Chem.* **2012**, *14*, 1515–1523. [CrossRef]
104. White, R.J.; Yoshizawa, N.; Antonietti, M.; Titirici, M.-M. A sustainable synthesis of nitrogen-doped carbon aerogels. *Green Chem.* **2011**, *13*, 2428. [CrossRef]
105. Braghiroli, F.L.; Fierro, V.; Izquierdo, M.T.; Parmentier, J.; Pizzi, A.; Celzard, A. Nitrogen-doped carbon materials produced from hydrothermally treated tannin. *Carbon* **2012**, *50*, 5411–5420. [CrossRef]
106. Braghiroli, F.L.; Fierro, V.; Parmentier, J.; Vidal, L.; Gadonneix, P.; Celzard, A. Hydrothermal carbons produced from tannin by modification of the reaction medium: Addition of H^+ and Ag^+. *Ind. Crops Prod.* **2015**, *77*, 364–374. [CrossRef]
107. Navarrete, P.; Pizzi, A.; Bertaud, F.; Rigolet, S. Condensed tannin reactivity inhibition by internal rearrangements: Detection by CP-MAS ^{13}C NMR. *Maderas Cienc. Tecnol.* **2011**, *13*, 59–68. [CrossRef]
108. Young, D.A.; Cronjé, A.; Botes, A.L.; Ferreira, D.; Roux, D.G. Synthesis of condensed tannins. Part 14. Biflavanoid profisetinidins as synthons. The Acid-induced 'phlobaphene' reaction. *J. Chem. Soc. Perkin Trans.* **1985**, *1*, 2521–2527. [CrossRef]
109. Braghiroli, F.L.; Fierro, V.; Izquierdo, M.T.; Parmentier, J.; Pizzi, A.; Delmotte, L.; Fioux, P.; Celzard, A. High surface—highly N-doped carbons from hydrothermally treated tannin. *Ind. Crops Prod.* **2015**, *66*, 282–290. [CrossRef]
110. Braghiroli, F.; Fierro, V.; Pizzi, A.; Rode, K.; Radke, W.; Delmotte, L.; Parmentier, J.; Celzard, A. Reaction of condensed tannins with ammonia. *Ind. Crops Prod.* **2013**, *44*, 330–335. [CrossRef]
111. Braghiroli, F.L.; Fierro, V.; Szczurek, A.; Stein, N.; Parmentier, J.; Celzard, A. Hydrothermally treated aminated tannin as precursor of N-doped carbon gels for supercapacitors. *Carbon* **2015**, *90*, 63–74. [CrossRef]
112. Deng, X.; Liu, X.; Duan, T.; Zhu, W.; Yi, Z.; Yao, W. Fabricating a graphene oxide—bayberry tannin sponge for effective radionuclide removal. *Mater. Res. Express* **2016**, *3*, 055002. [CrossRef]
113. Braghiroli, F.; Fierro, V.; Szczurek, A.; Gadonneix, P.; Ghanbaja, J.; Parmentier, J.; Medjahdi, G.; Celzard, A. Hydrothermal treatment of tannin: A route to porous metal oxides and metal/carbon hybrid materials. *Inorganics* **2017**, *5*, 7. [CrossRef]
114. Braghiroli, F.L. Polyphénols Végétaux Traités par Voie Humide: Synthèse de Carbones Biosourcés Hautement Poreux et Applications. Doctoral Dissertation, Université de Lorraine, Épinal, France, 2014.
115. Pandey, A.; Bhaskar, T.; Stöcker, M.; Sukumaran, R. *Recent Advances in Thermochemical Conversion of Biomass*; Elsevier: Oxford, UK, 2015.
116. Marsh, H.; Rodríguez-Reinoso, F. *Activated Carbon*, 1st ed.; Elsevier: Amsterdam, The Netherlands, 2006.
117. Marsh, H.; Rodríguez-Reinoso, F. Chapter 5—Activation Processes (Thermal or Physical). In *Activated Carbon*; Elsevier Science Ltd.: Oxford, UK, 2006; pp. 243–321.
118. Szczurek, A.; Amaral-Labat, G.; Fierro, V.; Pizzi, A.; Celzard, A. Chemical activation of tannin-based hydrogels by soaking in KOH and NaOH solutions. *Microporous Mesoporous Mater.* **2014**, *196*, 8–17. [CrossRef]
119. Marsh, H.; Rodríguez-Reinoso, F. Chapter 6—Activation Processes (Chemical). In *Activated Carbon*; Elsevier Science Ltd.: Oxford, UK, 2006; pp. 322–365.
120. Reimerink, W.M.T.M. The use of activated carbon as catalyst and catalyst carrier in industrial applications. In *Studies in Surface Science and Catalysis*; Elsevier: Amsterdam, The Netherlands, 1999; Volume 120, pp. 751–769.
121. Chen, J.Y. *Activated Carbon Fiber and Textiles*; Elsevier: Duxford, UK, 2017; pp. 3–20.
122. Wu, D.; Fu, R.; Sun, Z.; Yu, Z. Low-density organic and carbon aerogels from the sol–gel polymerization of phenol with formaldehyde. *J. Non-Cryst. Solids* **2005**, *351*, 915–921. [CrossRef]
123. Szczurek, A.; Amaral-Labat, G.; Fierro, V.; Pizzi, A.; Masson, E.; Celzard, A. Porosity of resorcinol-formaldehyde organic and carbon aerogels exchanged and dried with supercritical organic solvents. *Mater. Chem. Phys.* **2011**, *129*, 1221–1232. [CrossRef]
124. Rey-Raap, N.; Szczurek, A.; Fierro, V.; Menéndez, J.A.; Arenillas, A.; Celzard, A. Towards a feasible and scalable production of bio-xerogels. *J. Colloid Interface Sci.* **2015**, *456*, 138–144. [CrossRef]
125. Grishechko, L.I.; Amaral-Labat, G.; Fierro, V.; Szczurek, A.; Kuznetsov, B.N.; Celzard, A. Biosourced, highly porous, carbon xerogel microspheres. *RSC Adv.* **2016**, *6*, 65698–65708. [CrossRef]
126. Amaral-Labat, G.; Szczurek, A.; Fierro, V.; Stein, N.; Boulanger, C.; Pizzi, A.; Celzard, A. Pore structure and electrochemical performances of tannin-based carbon cryogels. *Biomass Bioenergy* **2012**, *39*, 274–282. [CrossRef]

127. Szczurek, A.; Fierro, V.; Medjahdi, G.; Celzard, A. Carbon aerogels prepared by autocondensation of flavonoid tannin. *Carbon Resour. Convers.* **2019**, *2*, 72–84. [CrossRef]
128. Sanchez-Sanchez, A.; Izquierdo, M.T.; Mathieu, S.; González-Álvarez, J.; Celzard, A.; Fierro, V. Outstanding electrochemical performance of highly N- and O-doped carbons derived from pine tannin. *Green Chem.* **2017**, *19*, 2653–2665. [CrossRef]
129. Fernández-Barbero, A.; Suárez, I.J.; Sierra-Martín, B.; Fernández-Nieves, A.; de las Nieves, F.J.; Marquez, M.; Rubio-Retama, J.; López-Cabarcos, E. Gels and microgels for nanotechnological applications. *Adv. Colloid Interface Sci.* **2009**, *147–148*, 88–108. [CrossRef]
130. Braghiroli, F.L.; Bouafif, H.; Neculita, C.M.; Koubaa, A. Activated biochar as an effective sorbent for organic and inorganic contaminants in water. *Water. Air. Soil Pollut.* **2018**, *229*, 230. [CrossRef]
131. Bi, Z.; Kong, Q.; Cao, Y.; Sun, G.; Su, F.; Wei, X.; Li, X.; Ahmad, A.; Xie, L.; Chen, C.-M. Biomass-derived porous carbon materials with different dimensions for supercapacitor electrodes: A review. *J. Mater. Chem. A* **2019**, *7*, 16028–16045. [CrossRef]
132. Frackowiak, E.; Béguin, F. Carbon materials for the electrochemical storage of energy in capacitors. *Carbon* **2001**, *39*, 937–950. [CrossRef]
133. Hsieh, C.-T.; Teng, H. Influence of oxygen treatment on electric double-layer capacitance of activated carbon fabrics. *Carbon* **2002**, *40*, 667–674. [CrossRef]
134. Wei, L.; Sevilla, M.; Fuertes, A.B.; Mokaya, R.; Yushin, G. Hydrothermal carbonization of abundant renewable natural organic chemicals for high-performance supercapacitor electrodes. *Adv. Energy Mater.* **2011**, *1*, 356–361. [CrossRef]
135. Zhao, L.; Fan, L.-Z.; Zhou, M.-Q.; Guan, H.; Qiao, S.; Antonietti, M.; Titirici, M.-M. Nitrogen-containing hydrothermal carbons with superior performance in supercapacitors. *Adv. Mater.* **2010**, *22*, 5202–5206. [CrossRef]
136. Si, W.; Zhou, J.; Zhang, S.; Li, S.; Xing, W.; Zhuo, S. Tunable N-doped or dual N, S-doped activated hydrothermal carbons derived from human hair and glucose for supercapacitor applications. *Electrochim. Acta* **2013**, *107*, 397–405. [CrossRef]
137. Zapata-Benabithe, Z.; Diossa, G.; Castro, C.D.; Quintana, G. Activated carbon bio-xerogels as electrodes for supercapacitors applications. *Procedia Eng.* **2016**, *148*, 18–24. [CrossRef]
138. Alvares Rodrigues, L.; Koibuchi Sakane, K.; Alves Nunes Simonetti, E.; Patrocínio Thim, G. Cr total removal in aqueous solution by PHENOTAN AP based tannin gel (TFC). *J. Environ. Chem. Eng.* **2015**, *3*, 725–733. [CrossRef]
139. Kunnambath, P.M.; Thirumalaisamy, S. Characterization and utilization of tannin extract for the selective adsorption of Ni(II) ions from water. *J. Chem.* **2015**, *2015*, 1–9. [CrossRef]
140. Nakajima, A. Electron spin resonance study on the vanadium adsorption by persimmon tannin gel. *Talanta* **2002**, *57*, 537–544. [CrossRef]
141. Nakajima, A.; Ohe, K.; Baba, Y.; Kijima, T. Mechanism of gold adsorption by persimmon tannin gel. *Anal. Sci. Int. J. Jpn. Soc. Anal. Chem.* **2003**, *19*, 1075–1077. [CrossRef] [PubMed]
142. Kim, Y.-H.; Alam, M.N.; Marutani, Y.; Ogata, T.; Morisada, S.; Nakano, Y. Improvement of Pd(II) adsorption performance of condensed-tannin gel by amine modification. *Chem. Lett.* **2009**, *38*, 956–957. [CrossRef]
143. Morisada, S.; Kim, Y.-H.; Ogata, T.; Marutani, Y.; Nakano, Y. Improved adsorption behaviors of amine-modified tannin gel for palladium and platinum ions in acidic chloride solutions. *Ind. Eng. Chem. Res.* **2011**, *50*, 1875–1880. [CrossRef]
144. Ogata, T.; Kim, Y.H.; Nakano, Y. Selective recovery process for gold utilizing a functional gel derived from natural condensed tannin. *J. Chem. Eng. Jpn.* **2007**, *40*, 270–274. [CrossRef]
145. Gurung, M.; Adhikari, B.B.; Kawakita, H.; Ohto, K.; Inoue, K.; Alam, S. Selective recovery of precious metals from acidic leach liquor of circuit boards of spent mobile phones using chemically modified Persimmon tannin gel. *Ind. Eng. Chem. Res.* **2012**, *51*, 11901–11913. [CrossRef]
146. Morisada, S.; Rin, T.; Ogata, T.; Kim, Y.-H.; Nakano, Y. Adsorption removal of boron in aqueous solutions by amine-modified tannin gel. *Water Res.* **2011**, *45*, 4028–4034. [CrossRef]
147. Ogata, T.; Morisada, S.; Oinuma, Y.; Seida, Y.; Nakano, Y. Preparation of adsorbent for phosphate recovery from aqueous solutions based on condensed tannin gel. *J. Hazard. Mater.* **2011**, *192*, 698–703. [CrossRef]

148. Sánchez-Martín, J.; González-Velasco, M.; Beltrán-Heredia, J.; Gragera-Carvajal, J.; Salguero-Fernández, J. Novel tannin-based adsorbent in removing cationic dye (Methylene Blue) from aqueous solution. Kinetics and equilibrium studies. *J. Hazard. Mater.* **2010**, *174*, 9–16. [CrossRef]
149. Rahman, M.; Akter, N.; Karim, M.R.; Ahmad, N.; Rahman, M.M.; Siddiquey, I.A.; Bahadur, N.M.; Hasnat, M.A. Optimization, kinetic and thermodynamic studies for removal of Brilliant Red (X-3B) using Tannin gel. *J. Environ. Chem. Eng.* **2014**, *2*, 76–83. [CrossRef]
150. Fayemiwo, O.M.; Daramola, M.O.; Moothi, K. Tannin-based adsorbents from green tea for removal of monoaromatic hydrocarbons in water: Preliminary investigations. *Chem. Eng. Commun.* **2018**, *205*, 549–556. [CrossRef]

© 2019 by the authors. Licensee MDPI, Basel, Switzerland. This article is an open access article distributed under the terms and conditions of the Creative Commons Attribution (CC BY) license (http://creativecommons.org/licenses/by/4.0/).

Review

Tannins: Prospectives and Actual Industrial Applications

Antonio Pizzi

LERMAB-ENSTIB, University of Lorraine, 27 rue Philippe Seguin, 88000 Epinal, France; antonio.pizzi@univ-lorraine.fr

Received: 31 May 2019; Accepted: 3 August 2019; Published: 5 August 2019

Abstract: The origin of tannins, their historical evolution, their different types, and their applications are described. Old and established applications are described, as well as the future applications which are being developed at present and that promise to have an industrial impact in the future. The chemistry of some of these applications is discussed where it is essential to understand the tannins and their derivates role. The essential points of each application, their drawbacks, and their chance of industrial application are briefly discussed. The article presents historical applications of tannins, such as leather, or traditional medicine, and more recent applications.

Keywords: tannins; applications; new applications; drawbacks; advantages

---

1. Origin of Tannins

Leather tanning has been used for centuries, even millennia, by immersing skins in water where special barks or woods containing tannin have been added. Up to a full year was necessary for leather to be produced in such a manner. However, the current tannin extraction industry is relatively newer. The more modern history of tannins began in the 17th century when Giovannetti, an Italian chemist, studied the interactions between iron solutions and substances called "astringents". In 1772, various researchers identified the presence of an acid in these compounds. This acid was then isolated by Scheele and turned out to be gallic acid. Based on the experiments of Deyeux and Bartholdi, continued by Proust in the late 18th and early 19th centuries, tannins have been officially recognized as a discrete group of different molecules based on gallic acid content. The great growth of the tannin extraction industry began in the years around 1850 in Lyon, where tannin was used as iron tannate for the black coloring of silk for women's blouses [1]. After 10 years, fashion changed and thus after many bankruptcies and groupings of factories, tannin manufacturers were able to convince the leather industry to use tannin in place of oak chips with considerable savings in tanning time (from 12 months with the old system based on wood chips rich in tannin to 28 days using tannin extract) [1]. The benefits of tannin extracts in the manufacture of leather, and even the time savings still allowed by their use, were such that the industry expanded rapidly and thrived. Tannin being in short supply in Europe, factories were opened in distant countries to satisfy the growing demand and promoting the use of alternative tannin types. Early in the 20th century, South American and southern and central African tannins began to be industrially extracted to supply major markets in Europe and North America. Among these, the main ones were quebracho wood and mimosa bark tannins. Leather processing was thus the second major boom period for the tannin industry. After World War II the substitution of leather with synthetic materials for shoes again caused a number of tannin extraction plants to close [2]. The third period of use of tannin therefore began, first with their development as bio-based adhesives and later with an increasing number of applications in new bio-based materials, this latter period being still in full development.

The name "tannin" comes from the use of this class of compounds in the tanning process of hides to give leather. Their general appearance varies, ranging from white amorphous powders to off-white amorphous powders, to glossy, almost colorless pasty substances, to reddish-brown powders when produced by spray drying. They have an astringent taste. Tannins are natural products found in most higher plants. They are produced in almost all parts of the plant, namely seeds, roots, bark, wood, and leaves, because of their fundamental role in the defense of the plant against insects, food infections, fungi, or bacteria. The defense mechanism is based on the ability of tannins to complex proteins irreversibly. They are also considered as one of the effective components contributing to the fact that the risk of suffering from cardiovascular diseases and some forms of cancer can be reduced by choosing diets rich in fruits and vegetables. In addition to their documented effects on human health, tannins are also important for the welfare of ruminants; high protein feeds such as alfalfa trigger the production in the rumen of methane trapped as proteinaceous foam, resulting in a potentially mortal fermentation that can be reduced by adding tannin in the diet. Two wide classes of tannins exist: hydrolysable tannins such as gallo-tannins and ellagi-tannins, and condensed polyflavonoid tannins, these latter being stable and rarely subject to hydrolysis [2].

1.1. Hydrolysable Tannins

The hydrolysable tannins, usually present in small amounts in plants, are simple derivatives of gallic acid, and they are classified according to the products obtained after hydrolysis—gallo-tannins (gallic acid compounds and glucose) and ellagi-tannins (composed of biaryl units and glucose).

Most gallo-tannins isolated from plants contain a polyol residue derived from D-glucose, although a large variety of polyol types can be found. Two other categories, gallo-tannins of tara (composed of gallic and quinic acid and glucose) and caffe-tannins (quinic acid esterified with caffeic acids plus glucose compounds), also occur [3,4] (Figure 1).

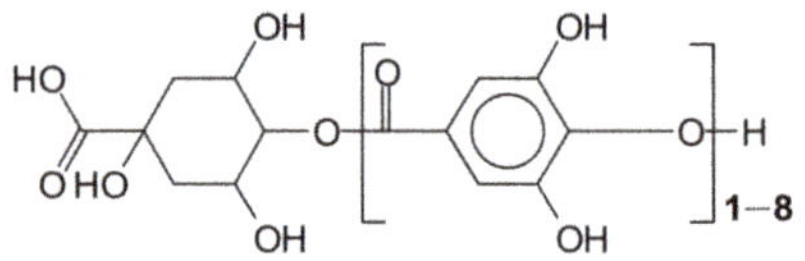

Figure 1. Example of the structures of tara tannin [3,4] and caffe-tannins (quinic acid esterified with caffeic acids plus glucose compounds).

1.1.1. Gallo-tannins

Tannic acid is one of the most important substances in relation to hydrolysable tannins. Tannic acid exists de facto in the form of a mixture of very similar substances, for example penta-(digalloyl)-glucose and tetra-(digalloyl)-glucose or tri-(digalloyl)-di-(galloyl)-glucose, etc. [4–6].

1.1.2. Ellagi-tannins

Different from gallo-tannins, ellagi-tannins contain additional binding motifs that arise from additional oxidative coupling reactions between the galloyl fragments [6]. The biosynthesis of ellagi-tannin is therefore an oxidative enzymatic progression of gallo-tannins. The first step is an oxidation of 1,2,3,4,6-pentagalloylglucose to form the monomeric ellagi-tannin. The second step consists essentially of dimerization after a second enzyme-mediated specific oxidation of hexahydroxydiphenoyl group of the ellagi-tannin with the galloyl group of another tannin to form a valoneyl group-containing dimer and these reactions continue to form higher oligomers. An example of ellagi-tannin is the tannin of chestnut wood (Figure 2).

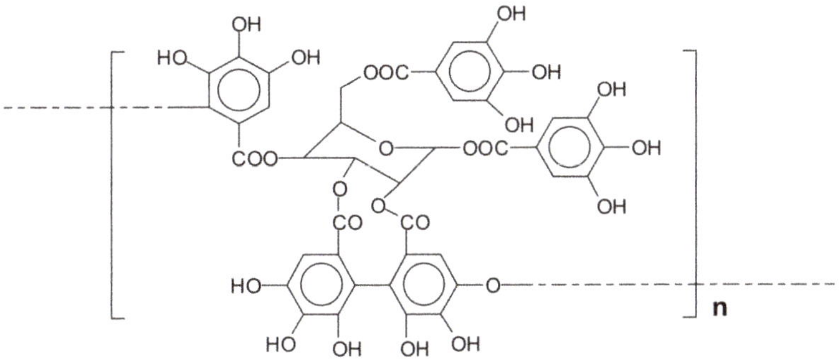

Figure 2. Schematic representation of the repeating unit of the ellagi-tannin of chestnut wood.

1.2. Condensed Polyflavonoid Tannins

Condensed tannin extracts consist of flavonoid oligomers of different average degrees of polymerisation. Small proportions of flavan-3-ols, flavan-3,4-diols, and other flavonoid analogs are also present [7,8]. Carbohydrates, such as broken down residues of hemicelluloses, and hexoses, pentoses, and disaccharides together with some imino acids and amino acids [2,9] constitute the non-phenolic part of tannin extracts. These latter, as well as the monoflavonoids, are equally present in too low a proportion to influence the extract properties. On the contrary, oligomers derived from hydrolysed hemicelluloses are often present in sufficient quantities. Equally, carbohydrate chains of various lengths [4,5] are also sometime linked to the flavonoid unit in the tannin. The basic structure of these tannins is based on the flavonoid unit (Figure 3).

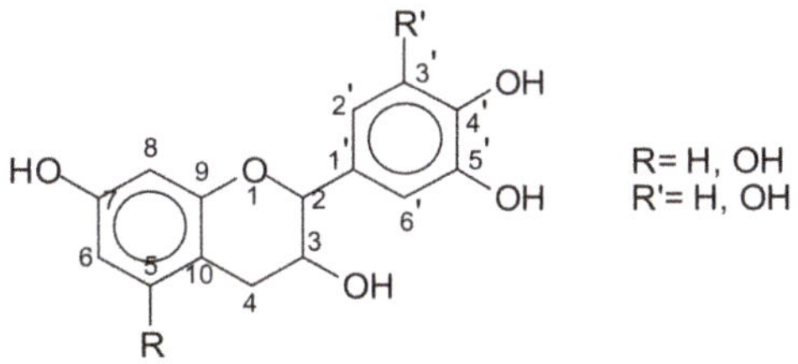

Figure 3. The structure of a flavonoid unit.

These flavonoid units are generally linked C4 to C6, or C4 to C8 to forms a variety of short chains of different lengths according to the type of tannin.

2. Industrial Utilization

The industrial uses briefly described below are in order of present or probable future importance to give an idea of what is developed and used already, and which applications are likely to gather importance in the future.

2.1. Leather Tanning

The manufacture of leather is still the largest use of tannins of vegetable origin. Leather has traditionally been made in ground pits in which alternating layers of animal skins and wood chips containing tannins, such as oak chips, have been placed and soaked for considerable periods of time. The skins were passed through a number of consecutive pits, generally 28, so as to be slowly enriched and "tanned" by the tannin in solution. The tannins exfiltrated the wood chips and slowly impregnated increasingly more of each skin. Such a manufacturing system was practiced for many centuries and produced good quality leather but it took several months, often a whole year, for the leather to be ready. The first change came when the tannin extraction industry, which had grown considerably in the 1850s to supply black iron tannate dyes to color silk, was in a desperate position because of the change in fashion. Some tannin extraction plants were able to demonstrate that by directly adding tannin extract to traditional tanning pits, the same quality of leather could be produced in just 28 days (28 pits, one per day). Leather tanned in this way began to prevail and this use led to a boom which lasted mainly until the end of the Second World War. In particular, the war years were good simply because armies were walking in leather boots.

At the end of the Second World War, three events contributed significantly to the steady decline in vegetable tanned leather. First, the introduction of synthetic materials, derived from petrol, for shoe soles begun to compete strongly with leather for one of its most traditional applications. Secondly, the demobilization of armies, which sharply reduced the need for leather shoes, and the third, the strong penetration of the market by chromium salt tanning for the manufacture of soft leathers, in particular the upper part of shoes. With all these changes, while vegetable leather has now begun to gain a reputation as a luxury product, there are still some important niches for which it is used such as equestrian equipment, heavy bags and luggage, and other heavy applications as well as the real high price luxury markets. Although the traditional 28 pits tanning method still exists for a number of special leathers as well as in the case of artisanal leather (as in Morocco for example), the tanning processes has also evolved for vegetable tanning where rotary drums, a technique borrowed from chrome tanning, allows vegetable tanning to be finalized in about 24 h. Today, research on vegetable tanned leather has been able to produce much more supple leather through the inclusion of oils and other techniques, so that some rebirth of the use of plant tannins for other application areas appears to take place.

2.2. Wood Adhesives

There are a number of detailed reviews on the use of tannins for wood adhesives. The reader is referred to these detailed studies [2,10]. However, here existing technologies and industrial use of wood tannin adhesives are presented.

As extensive studies already exist, and this application of tannin is now the second most important after leather manufacturing, only a few of the main achievements of tannin-based adhesives for wood products will be highlighted. (1) The development, optimization, and industrialization of non-fortified but chemically modified thermosetting tannins for particleboard, other particle products, and plywood [11–13]. (2) The technology for rapidly pressing tannin adhesives for particle board, which is also industrial [14,15]. (3) The development and industrialization of tannin–urea–formaldehyde adhesives for plywood and in particular as impregnators for corrugated board starch binders [16]. (4) The development and industrialization of cold-setting tannin–resorcinol–formaldehyde adhesives for glulam and fingerjointing [17]. (5) The large-scale development and industrialization of fast-setting "honeymoon" separate application cold-setting adhesives for tannin-bonded glulam and fingerjoints [18–20] (Figure 4).

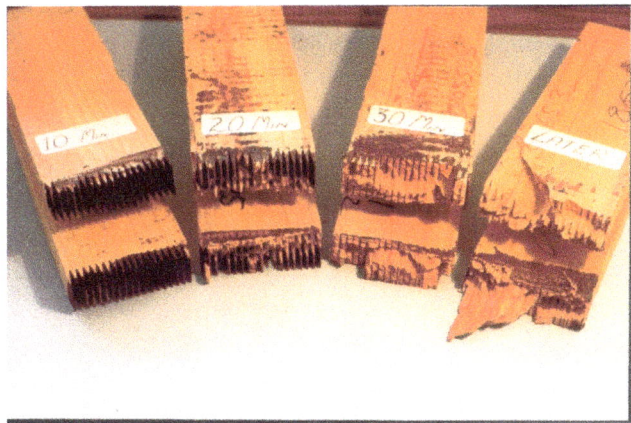

Figure 4. Development of strength visualized through the rapid increase in percentage wood failure of tannin-based fast-setting "honeymoon" separate application cold-setting adhesives for glulam and fingerjoints.

(6) The development and industrialization of zinc salts to accelerate the hardening of non-fortified tannin adhesives for plywood [2,21–23]. (7) Successful formulation, development, and industrialization in Chile of pine bark tannin adhesives for particle boards and for glulam and fingerjointing [13,24]. (8) The development of isocyanate/tannin copolymers as difficult-to-bond hardwood adhesives and for plywood and other applications [25,26]. (9) The development of very low formaldehyde tannin adhesives for particle boards and other wood panels. (10) The development of the use of hardeners other than formaldehyde for thermosetting tannin adhesives [2,27,28]. (11) The discovery and development of self-condensation of tannin for adhesives [29–36].

All industrialized technologies today are based on paraformaldehyde or hexamethylene tetramine (hexamine) [36]. The latter is much more user and environmentally friendly.

As regards wood adhesives, a number of experimental improvements have been studied, dictated by the new environment in which wood adhesives must operate. First of all, the relative scarcity of tannins produced in the world, compared to the tonnage of synthetic adhesives used in the panel industry, has led to a great deal of research on the extension of the tannin resource in order to have larger tonnage. As the potential material for tannin extraction shows that millions of tons of this material can be extracted each year worldwide, some companies have started to build additional extraction plants. This movement is still relatively small, but it is ongoing. The second approach, to extend the tannin with another abundant and natural material, has led to the preparation of adhesives based on in situ copolymers of tannins and lignin [37] or copolymers of tannin and protein or soy flour [38], and the use of tannin–furfuryl alcohol adhesive formulations, furfuryl alcohol being also a bio-based material [39].

The second new constraint is the demand of most companies to eliminate formaldehyde emissions from tannin adhesive. This quest has taken two approaches: (1) total elimination of formaldehyde by substituting it with aldehydes, which are less or non-toxic, and non-volatile [28,40], such as glyoxal, glutaraldehyde, or vanillin, the latter giving a fully bio-based tannin adhesive, and even aldehydes generated by the action of sodium periodate on glucose, sucrose and even oligomeric carbohydrates, (2) the use of non-aldehyde hardeners such as trishydroxymethylnitromethane [41] and trishydroxymethylaminomethane [42] or even by combination with furfuryl alcohol, the latter functioning both as a hardener and a contributor to a tannin/furan copolymer [39,43]. (3) The use of hexamine with the formation of –CH_2–NH–CH_2– bridges between the tannin molecules, where the secondary amine is capable of absorbing any emission of formaldehyde from the heating of the wood itself or any other emission of formaldehyde to produce truly zero-formaldehyde emission panels [36,44–46]. (4) Lastly, the hardening of the tannins by autocondensation without the addition of a hardener, autocondensation catalysed by the wood substrate itself in the case of fast-acting procyanidin tannins, such as pine bark tannins, and for slower tannins by addition of silica or silicate or other accelerators [10,29–35] allowing the preparation of wood particleboard of indoor quality.

2.3. Pharmaceutical and Medical Applications

Tannins are known bactericides because they react with proteins irreversibly, thus complexing within bacterial membranes, neutralizing their activity. As a consequence, tannin-based pharmaceuticals to cure intestine infections have long-time been marketed. They have effective anticaries properties. Tannins have also many applications for other pharmaceutical/medical uses but all these are targeted for future use rather than the present.

Several experimental studies on the pharmaceutical use of tannins have been published with antitumor and anti-oncogenic activities particularly well documented [47–52]. Their antiviral effectiveness is also well documented by in vitro screening for a variety of 12 different hydrolysable and condensed tannins [51]. The tannin's minimal inhibitory concentration (MIC) needed for reducing by 50% the cytopathogenicity induced by a number of viruses was used as an evaluation of their effectiveness. The lower MIC values yielded the best antiviral behavior. The different tannin's minimum cytotoxic concentration (MCC) needed to detect microscopic alteration of normal cell morphology was also determined. Less toxic is the tannin tested in the patient's cells, thus the higher is its MCC value, the more acceptable it is as an antiviral compound. The ideal antiviral compound is then the one presenting a combination of the lowest MIC and the highest MCC. The effectiveness of different tannins due to their polyphenolic nature can be very high against a number of different viruses. This is due to their irreversible reaction and combination with the viruses capsid proteins. It is the same reaction used in leather tannins and in their association with carbohydrates.

Thus, a number of commercially available tannins have been tested, namely mimosa bark tannin extract and its derivatives, chestnut tannin extract, tara tannin, quebracho wood tannin extract both sulphited and natural, pecan nut tannin extract, pine bark tannin extract, sumach tannin extract, and spruce tannin extract [51]. The viruses against which all these have been tested are highly varied, such as HIV-1 and HIV2, Herpes simplex 1 and 2, Vaccinia virus, vesicular stomatitis virus, Coxsackie virus B4, respiratory syncytial virus, Influenza A H1-N1, Influenza A H3-N2, Influenza B, Human Corona virus, Reovirus-1, Feline Corona virus, Sindbis virus, para-influenza 3 virus, and Punta Toro virus [52].

The inhibitory effects of these tannins have also been tested on proliferation of murine leukemia cells, murine mammary carcinoma cells, and human T-cells [51].

Acutissimin A is a bound flavonoid with an ellagi-tannin. It is formed by the interaction of a wine flavonoid with the vescalagin generated by the barrel's oak tannin [53,54]. Acutissimin A has been found to present an effectiveness 250 times higher to stop tumors growth than the drug Etoposide.

While many studies have been conducted on a variety of tannins derived from a wide variety of plants as an anticancer treatment, some studies on the possibility of using tannin for other medical applications have also been highlighted. Condensed tannins are traditionally used for the treatment of intestinal problems [55,56]. This is due to their complexation ability with other molecules and their antioxidant behavior. The extract of *Stryphnodendron rotundifolium* and of other tannins has proven their effectiveness against ulcers by functioning as a protective coating of the gastrointestinal tract [57–59]. Other possible mechanisms of action of phenolic plant extracts as herbal medicines against ulcers and gastritis have also been described [57–60].

2.4. Wine, Beer, and Fruit Juices Additives and Antioxidants

Wine, beer, and fruit juices naturally contain tannins [61]. It is actually their presence that accounts for their characteristic taste. In short, the level of tannin in any of these products must be within a definite interval/concentration range for the beverage to be organoleptically pleasing. Too low an amount of tannin and the beverage will be insipid and with no taste. Too high the proportion of tannins in the beverage and it is too unpleasant, too "tanning" for the consumer's mouth. Many wines, some beers, and several fruit juices however contain too low a concentration of tannin and thus may need to be "doctored". Initially, addition of tannin or tannin-rich wood chips directly to wine to enhance its taste and give the impression of a wine of greater age was strictly forbidden in most European countries. With the determined and successful push for wine markets by southern hemisphere producers where limitations on adding oak particles to the wine to accelerate its aging was not forbidden, producers of more established countries tried to defend their market in a different way. Some producers the wines of which were particularly low in tannin content started to add tannin directly to some of their wines. Although addition of additives for aging was legally forbidden the perfectly legal gap existed permitting the use of tannins to precipitate proteic matter in the wine to render it "clearer". All what was needed then was to add more tannins than what would be required to render the wine "clear". This was kept fairly confidential, to not incur potential problems. The situation changed dramatically once the so-called "French paradox" came to be known. Namely, notwithstanding that traditional French diet is very rich in fats that should lead to grave cardiovascular diseases this is not the case, and this type of disease was far less frequent in France. This was ascribed to the regular use of red wine which decreased to extremely low level the risks and the occurrence of cardiovascular diseases. Now not only is tannin added to the wine, but it is considered particularly beneficial to do it. It must be pointed out that it is particularly purified tannin, from which carbohydrates and other components have been eliminated. It is the antioxidant property of the polyphenolic groups of the tannin which gives to it the powerful anticardiovascular effect and positive properties. Both addition of tannin as well as addition of tannin-rich oak wood chips to wine is now completely legal in Europe.

Equally, some beers need addition of purified hydrolysable tannins to acquire the proper taste. The same is valid for fruit juices.

As a consequence of the public awareness generated by the "French paradox" diet complements rich in flavonoid and hydrolysable tannins are on sale "over the counter" in North America.

This antioxidant effect of tannin is now starting to be exploited for foodstuff. The antioxidant effect of tannins is due to the characteristic of any phenol to stabilize free radicals and to inhibit further damage these may otherwise cause. Several factors influence radical control, these being [58] (1) the tannin colloidal state in water, (2) the special conformation around the interflavonoid bond, (3) how easy it is to open the pyran C-ring heterocycle, (4) the relative A- and B-ring hydroxygroups proportions, and (5) how easy is to solvate the tannin. These are the parameters determining the capacity of a tannin to act as an antioxidant.

A tannin's antioxidant capability can be defined by measuring two different variables:

(1) How fast a tannin can form a radical, and uptake a radical by transfer from another radical species. Thus, the ease and ability of a tannin to capture free radicals from another species. The antioxidant ability is greater the easier and faster the radical capture.

(2) How quickly a tannin phenoxyl radical decays, thus how stable the radical is as a function of time. The slower the radical decay the lesser the radical degradation of the material to which tannin is added, hence the better the antioxidant properties of the tannin.

2.5. Fireproof and Insulating Foams

There are a number of different developments in rigid foam insulation. For imitating synthetic polyurethanes, by reaction with isocyanates, two approaches have been tried. First, foams have been developed based on the reaction of a modified tannin either by benzoylation or by oxypropylation to make it more susceptible as polyol to the reaction with polymeric isocyanates [62–65]. This first approach follows the same approach that was made with another natural polyphenol, namely lignin [66]. This is a traditional approach where tannin only functions as a polyol. The second approach of this type uses a very different strategy based on the reaction of a tannin with an aldehyde and the subsequent reaction of the methylol (–CH_2OH) groups so formed with the isocyanate. This approach is now used for both tannin wood adhesives [25,67] and other formaldehyde adhesives such as urea-formaldehyde

(UF), melamine-urea-formaldehyde (MUF) and phenol-formaldehyde (PF) [25,26]. Incidentally it is the only system that allows the formation of urethanes in an aqueous medium.

Thus, according to this second approach, mixed rigid foams of phenolic–polyurethane type have been developed by reacting natural tannin/furanic mixtures with polymeric isocyanate in the proportions suitable for continuous polyurethane foam lines [68]. Urethane linkages formed at both the alcoholic C3 and phenolic hydroxyls. Other species in the mix were involved in urethane linkages formation: (1) glyoxal after or before its reaction with the tannin, (2) the phenolsulfonic acid catalyst, and (3) furfural. Furfural instead preferred to form methylene bridges with the flavonoids A ring than to form urethane linkages by reacting with the isocyanate. Thus, methylene and urethane bridges were formed between all the main materials in the mix. Thus, a number of mixed species bound by both types of linkages were formed. Species formed by mixed co-reaction of two, three, and four different reagents in the mix were identified. Examples of mixed species are shown in Figure 5:

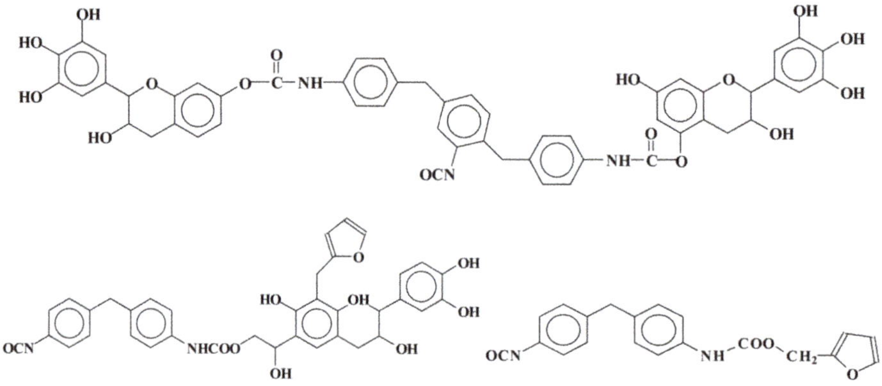

Figure 5. Examples of mixed species obtained by the reaction of tannin–furfuryl alcohol–glyoxal mixes with polymeric diphenyl methane isocyanate (PMDI).

Testing on a continuous production line has yielded positive results. This system was tried in plant trials in Switzerland on a continuous production line for polyurethane foam mats where addition of a small amount of isocyanate was necessary as without it the equipment of the factory could not function [68].

Chemically self-expanding rigid foam formulations based on tannin extracts have been developed since 1994 [69]. These foams, composed of 95% natural materials, have mechanical and physical properties comparable to synthetic PF foams. Originally, the fluid phase before foaming is composed of a tannin, with formaldehyde as a hardener, both mixed with furfuryl alcohol used as a exothermic reaction agent by its self-polymerization reaction and its reaction with the tannin [69] under acid conditions. Expansion to a foam of the fluid phase is caused by a low-boiling physical blowing agent, while simultaneous cross-linking of the resinous mixture provides dimensional stabilization at the desired low density [69]. These foams have been tested and are totally fire resistant, this being their major interest. Their appearance is shown in Figure 6.

Figure 6. Scanning electron microscope image of the structure of a first-generation tannin–furanic foam (**left**) and macro-appearance of the same (**right**).

During the last ten years a great deal of research has been carried out to improve these foams, by formulating them without formaldehyde [70,71], without aldehydes, without an organic solvent as blowing agent [72], without furfuryl alcohol, in an alkaline medium [73] and not only acid, with open cells and closed cells, by copolymerizing them with synthetic resins such as phenolic resins [74] and isocyanates [75], by copolymerizing them with proteins [76], by introducing variations into their systems of preparation [77,78], by improving their resistance to water by grafting on a small proportion of the tannin of long hydrorepellent chains [79], and also foaming them by heat alone without self-blowing, or even simply foamed by mechanical agitation for use as projected foams [80,81]. In addition to rigid foams, flexible foams have been developed [82–84]. Foams with the most reactive tannins of the procyanidin type, such as pine tannin, were also developed [85,86] and the carbonization line of research of these foams has also been continued and improved [87]. The varieties of these foams are really considerable [88] and the literature on this subject is really quite large, too vast to be summarized in this article.

A number of applications have been developed for these foams. Among others, open-cell foams have been used for very good acoustic insulation [89], carbonized foams for a large number of different applications [87,90], and also more recently for bone repair by osteogenesis with stem cells [91]. Taking the latter case as an example, for this application, as the treatment options are limited, bone tissue engineering opens the possibility of growing an unlimited amount of tissue products with increased therapeutic potential for the management of clinical cases characterized by severe bone loss. Bone engineering relies on the use of conformal biomaterial scaffolds, osteocompetent cells, and biologically active agents. Among other things, porous tannin spray dried powder (PTSDP) has been approved for human use. Thus the powder derived by grinding tannin–furanic foams has been used as a low-cost reconstruction material, due to its biocompatibility and potential ability to be replaced by bone in vivo. In this study, macro-PTSDP scaffolds with defined geometry, porosity, and mechanical properties were made by combining casting technology and pore leaching by mixing PTSDP and hydroxyapatite ($Ca_{10}(PO_4)_6(OH)_2$). This has shown that the scaffolds developed in this work support the attachment, long-term viability, and osteogenic differentiation of mesenchymal progenitors derived from human induced pluripotent stem cells. The combination of some macroporous PTSDP scaffolds with patient-specific osteocompetent cells offers therefore new opportunities to develop bone grafts with increased clinical potential for complex skeletal reconstructions.

The variety of uses of these foams is potentially such that many other applications can be envisaged, not only envisaged at present but to be developed in the future.

2.6. Calcite Depressant in Ore Flotation and Other Mining Applications

Unmodified mimosa bark and quebracho wood tannin extracts are used as depressants for unwanted calcite for the recovery of fluorspar in South Africa by a single mining firm. Consumption for such an application is relatively low at about 100 tons of tannin extract per month, and the extract is applied at the rate of 1 kg tannin per ton of low grade (20%–25% CaF_2) fluorspar ore [92].

Equally low is the consumption of tannin extracts for the separation of germanium from copper in the big open air copper mine in Chile. Due to the low percentage of the very expensive and rare gemanium in the copper ore, this is treated with flavonoid tannins, the B-rings of which complex preferentially with the germanium, allowing its separation. The tannin–germanium complex is then burned and the germanium recovered.

2.7. Flocculants and Precipitation of Polluting Materials by Complexation of Heavy Metals

Scavenger behavior of natural flavonoids for several kinds of metal ions is well-known in literature [93–96]. On the other hand, nowadays, the problem of purification of wastewater by heavy metal ions is more and more considered because there is a growing sensitivity around their environmental pollution [97,98].

The idea of using a tannin in powder form, or as a rigid foams to purify industrial wastewater, connects the natural ability of polyphenols to entrap metal ions with the easy removal system of such an innovative product [99–103].

A reliable proportionality has been found between initial concentration and percentage of metal ion adsorbed. The most important aspect to consider is that tannins are able to adsorb copper (12.5%) and lead (20.1%) ions in their structures [102]. Moreover, analysis results show that the adsorbing materials are not exhausted and it is possible to preview a scavenger activity even for more polluted water solution. More recently, *Pinus pinaster* tannin extracts have been used for the complexation and precipitation of antimony Sb ions [104].

Tannin extracts have also been used as flocculants for clay suspension in water treatment. Strongly acid conditions are used to link ethanolamine with mimosa tannins which are water soluble (Figure 7). These anpho-tannins have been used in South Africa for over 15 years as flocculants for clay suspensions in municipal water treatment. In this process no residual salts or ions remain in the treated water, the anpho-tannins combining with the suspended clay and coprecipitating [92].

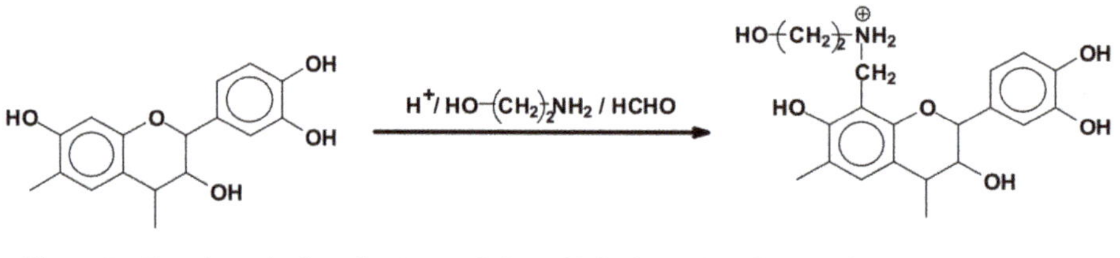

Figure 7. Reaction of ethanolamine and formaldehyde with a flavonoid tannin unit to form anpho-tannins used as industrial flocculants.

More recently, research has continued on the use of tannins as coagulants of particles in polluted waters to eliminate color even in wastewater by modifying the tannin by the Mannich reaction [105].

2.8. Inhibitors of Corrosion of Metals

Chemical and electrochemical acid corrosion of iron, steel, and alloy are common place. This causes great losses because the metal parts affected have to be replaced. Thus, anticorrosion coatings are a practical way of prevention. Metal surface preparation is essential when using such methods of corrosion prevention. An anticorrosive primer added before the main paint will markedly improve the protection of the substrate imparted by the paint finish. Many researchers have studied the anticorrosive properties of tannins. Tannin-based primers for the protection of metal surfaces and as anti-corrosion agents were already available on the market in the 1950s.

An anticorrosion primer formulation for steel has been published by Matamala et al. [106], this being primary layers of pine bark tannins that could prolong the paint life by more than 50% if applied before the main paint. In addition, tannins can be used as anticorrosives together with specific solvents and added to other materials such as epoxies, zinc oxide, or copper. This improves the quality of tannins to the same level of traditional anticorrosive primers. Such primers are applied by brushing or other suitable methods. Both hydrolyzable and condensed tannins can function as anticorrosive metal primers, as they both can potentially oxidize phenolic groups (antioxidant) and complex the metal substrate by forming of orthodiphenol metal complexes.

The reason for the coupling of tannins with metals is, for example, the formation of Fe complexes, used to prepare intensely black/violet inks by formation of ferric tannates, and also as anticorrosion varnishes. These coordination complexes are due, for example, to the orthodiphenol hydroxyl groups on the B rings of a flavonoid tannin [107] (Figure 8).

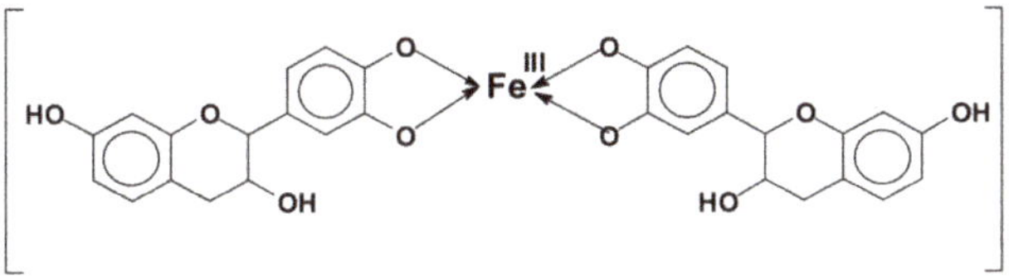

Figure 8. Example of orthodiphenol iron complex of a flavonoid tannin.

Ship hulls are traditionally coated with metal-containing anti-fouling paints. These provide protection by releasing a toxic compound. Tannins and copper can be used as antifouling for ship parts in seawater and reduce foulings. In addition, such formulations were able to reduce copper consumption by 40-fold compared to copper-based paints. It is this same type of complexation that has given the ferrous tannate-based inks still used in school in the first half of the 20th century and the use of which had its peak in the middle ages (cf. inks).

2.9. Core Binders for Foundry Sands

Foundry cores of high strength have been produced using both hot and cold-set tannin–formaldehyde adhesives to bind mold sands [92]. The mimosa or quebracho tannin–resol, which may be prepared and stored in solid form, reacts with more tannin extract (when mixed in the proportion 1:3 under heating and stoving at a temperature up to 170 °C). In cold-setting applications, setting of the resol with resorcinol may be acid catalyzed with ammonium chloride and p-toluene sulfonic acid. Setting of the cores from the cold-setting mix can be accelerated at any time by stoving.

2.10. Mud Stabilizers and Drilling Fluids

Drilling fluids play important roles in the drilling, such as in geothermal drilling programs, and serves several purposes in the circulation system. Among those are (1) cooling and lubricating the drilling strings while it is circulating, (2) transporting the cuttings that are created at the bottom of the hole and release them at the surface, (3) controlling the formation pressure while maintaining the wellbore stability, (4) sealing the permeable formations across the wellbore, and (5) assisting the logging operations. In drilling operations, the basic properties of drilling fluids are usually defined by the well program and monitored including rheology, density, fluid loss, solid content, and chemical properties. To avoid any drilling problems such as a blowout, stuck pipe, and other problems, engineers are always searching for a good performance of a drilling fluid system that is fit for the purpose. Dealing with standard temperature and pressure conditions, the properties of drilling fluids are much easier to handle. While in hostile drilling conditions like high temperature environments, the selection of a drilling fluid is more crucial, alongside the lithology that is one of the factors to be considered. The high temperature environment is often said to be at a temperature more than 150 °C. In geothermal drilling, the down hole temperature can reach up to 250–350 °C at the depth of 1000 to 2000 m. At this temperature, the selection of a drilling fluid becomes more stringent due to the needs of thermal stability of the drilling fluid system as well as materials used in the system. High temperatures can cause chemical alterations of various drilling fluid components, thus affecting the filtration and viscosity as well as increasing the tendency for gelation. In addition to this, clay activities greatly affect the system, with high tendencies to flocculate and gel.

Sulfited quebracho tannin extract has been traditionally used as a mud-thinning agent in shallow drilling [92]. However, stricter criteria of stability and anticorrosive action at elevated temperatures have been introduced in recent years. Thus, an effective conditioning agent for bentonite mud has been developed by incorporation of chrome salts into mimosa tannin extract [92]. This product is effective to a depth of 2000 m under normal drilling conditions and in absence of high salinity.

Tannin–lignosulfonate drilling additives that have been coreacted have also been prepared [108]. Calcium lignosulfonate (CaLS) is dissolved in an aqueous solution of hydrochloric acid for decalcification. The amount of CaLS added to the solution depends on the weight ratio between CaLS and the desired tannin. The mixture is frequently stirred for 10 min and sodium hydroxide is then added to adjust the pH of the solution to between 3 and 4. Then, the calcium sulfate is removed from the solution by filtration to obtain lignosulfonic acid. On a hot plate at a temperature of 105 °C, formaldehyde is added to the solution to initiate crosslinking. The tannin is then added according to the desired weight, and the solution is stirred continuously for 2 h. The ferrous sulfate heptahydrate is chelated with the solution for 30 min at the same temperature. Finally, the tannin–lignosulfonate solution is dried in a vacuum oven at a temperature of 60 °C for 48 h. The evaluation of the decomposition rate by thermogravimetric analysis (TGA) provided proof of the thermal stability of the tannin–lignosulfonate. The thermal stability of tannin–lignosulfonate has been identified at 294 °C and more than 60% of the weight remains at this temperature. It was found that the gel strength at 10 s and 10 min was reduced with the addition of deflocculant in each sample. [108].

2.11. Teflon/Metal Adhesives Resistant to High Temperatures and High Temperature Resistant Surface Finishes for Metals

The condensation and crosslinking reaction of mimosa tannin extract and a flavonoid monomer model with triethyl phosphate (PET) has been studied [109] (Figure 7). The reaction was shown by multiple instrumental analytical techniques to occur at the C3 of the flavonoid heterocycle and the flavonoid B-ring's C4 and C5 aromatic carbons, but not on the flavonoid A-ring. The relative proportions of reaction on these two sites differed for tannin and monomer model compounds. The reactions is temperature-dependent. The reaction, which takes place at temperatures around 180 °C, leads to hard solid finishes and films particularly suitable for attachment to aluminum and stainless steel and presents high thermal stability. Their potential use for which they were initially developed is in particular for the attachment of teflon coatings on non-stick steel or aluminum frying pans [109,110]. Subsequent testing with either of these tannin-based adhesives or other natural polyphenols [111] has demonstrated resistance to temperatures in excess of 400 °C for certain periods of time. Ammonia and high temperatures favor these reactions. The first application tests performed at high temperatures showed good performance as a metal coating. The type of polymers that are formed are as shown in Figure 9.

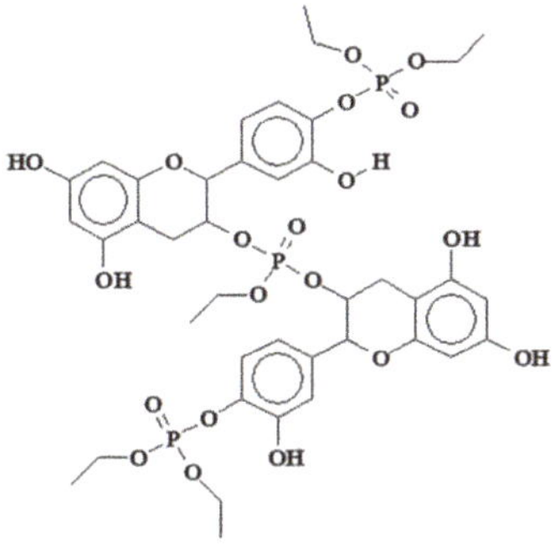

Figure 9. Example of cross-linking product of a flavonoid tannin by reaction with triethyl phosphate [109].

2.12. Cut Flowers and Hydroponic Horticulture Foams

High water absorption and retention, good air access, lower density for easy flower insertion, and particularly optimal pH to ensure durability are the required properties for synthetic phenol–furan synthetic foams for floral, hydroponic, and horticultural uses [112–115].

New quebracho tannin–furfuryl alcohol foams without formaldehyde have also been developed for the same applications [116] (Figure 8). Compounds to neutralize the inherent residual acidity of the catalyst used as a wetting agent are always included. Densities were in the 0.048 to 0.066 g/cm^3 range and compression strengths in the 0.07 to 0.09 MPa range. They present open pores with average cell sizes in the 125 to 250 μm range, peak water absorption up to 98% by volume, and a residual pH of 5. They do not present any phytotoxicity for preserving freshly cut flowers and are good supports for horticultural hydroponic cultures [116] (Figure 10).

Figure 10. Appearance of cut flowers (Transvaal daisies, *Gerbera* spp) on a tannin–furanic foam (black) compared to a synthetic phenol–formaldehyde/furanic foam (green).

Vermiculite and dolomite are added as well. The foams' pH determines their performance for rendering water and air available to the plants. The amount and structure of the anionic surfactant used also determines the water and air availability of the foam [116]. The inclusion of antifungal compounds and nutrients in the foam composition is also beneficial. Tannin foams based on renewable natural materials not only are environment-friendly but showed performance comparable to or better than the commercial synthetic phenolic floral foam used as a reference [116].

2.13. Corrugated Cardboard Adhesives

The adhesives developed for the manufacture of damp ply resistant corrugated cardboard are based on the addition of spray-dried wattle extract, urea-formaldehyde resin, and formaldehyde to a typical Stein–Hall starch formula of 18 to 22% starch content [16,117]. The wattle tannin–urea–formaldehyde copolymer formed in situ, and any free formaldehyde left in the glue line was absorbed by the wattle tannin extract. The wattle extract powder should be added at level of 4 to 5% of the total starch content of the mix (i.e., carrier plus slurry). Successful results can be achieved in the range of 2 to 12% of the total starch content, but 4% is the recommended starting level. The final level is determined by the degree of water hardness and desired bond quality. This wattle extract–UF fortifier system is highly flexible and can be adopted to damp-proof a multitude of basic starch formulations. This system has been in operation for some decades in countries where tannin is easily obtainable.

2.14. Non Isocyanate Polyurethanes (NIPU)

A high yield of urethane linkages was obtained by reacting hydroxyl-rich chestnut hydrolysable tannin with dimethyl carbonate followed by hexamethylenediamine [118]. Polyurethanes prepared in this way are of interest as toxic isocyanates and are not used as reagents, and their main constituent is instead a renewable material.

The same approach has been applied to condensed tannins. Several condensed tannin extracts were used for this purpose, namely: the bark tannins of maritime pine (*Pinus pinaster*), of mimosa (*Acacia mearnsii*), and radiata pine (*Pinus radiata*), and the wood tannins of quebracho (*Schinopsis lorentzii and balansae*). These were initially reacted with dimethyl carbonate followed by hexamethylenediamine forming urethane bonds [119]. Wood surface finishes based on these poly(hydroxyl)urethanes yielded encouraging performances (Figure 11) with high sessile drop contact angles on the treated wood surface (Figure 9).

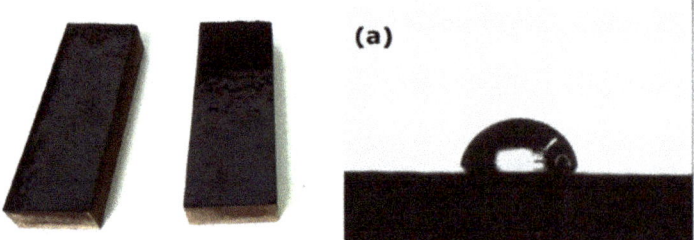

Figure 11. Non-isocyanate polyurethane-based (NIPU) wood surface coatings based on condensed tannins (**left**) and sessile water drop contact angle on the NIPU coating surface (**right**).

Furthermore, to increase considerably the proportion of bio-sourced material in these isocyanate-free polyurethanes, pre-aminated mimosa tannin extract (as a substitute of the synthetic hexamethylene diamine) was reacted with a mimosa tannin extract pre-reacted with dimethyl carbonate [120] (Figure 12). This reaction proceeded easily at room temperature. Cured above 100 °C, the polyurethanes formed yielded a hard film [120]. In addition, not only the polyphenolic fraction of the tannin extract, but also the carbohydrates oligomers in it generated isocyanate-free polyurethane linkages with the aminated tannin [118,120], indicating that non-isocyanate polyurethanes can be synthesized just from carbohydrates. The compounds formed were identified by several instrumental analytical techniques. Species as shown in Figure 12 were identified.

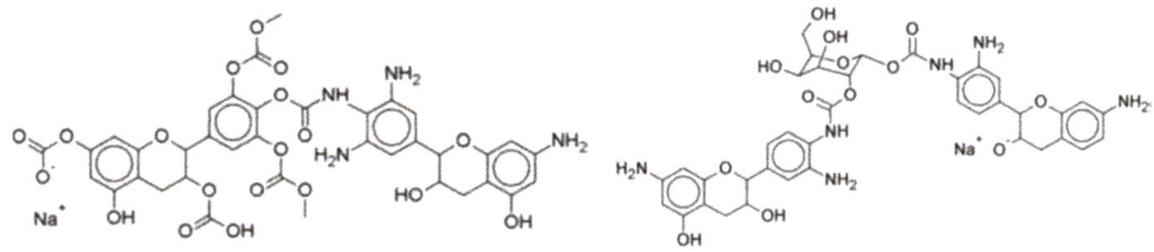

Figure 12. Examples of compounds presenting urethane linkages formed by the reaction of a pre-aminated flavonoid tannin with a flavonoid tannin pre-reacted with dimethyl carbonate.

The reactivity of the carbohydrates present either in the condensed tannins or in the hydrolysable tannin and the isolation of polyurethanes without isocyanates (NIPU) formed with the carbohydrates or carbohydrate–tannin mixtures led to the evolution of simple monosaccharides and disaccharides as a base of NIPU [121]. Thus, predominantly bio-sourced isocyanate-free polyurethanes (NIPUs) were developed from glucose and sucrose by reaction with dimethyl carbonate and hexamethylenediamine [121]. Oligomers obtained were detected by several spectrometry techniques, showing linear and branched structures such as those shown in Figure 13.

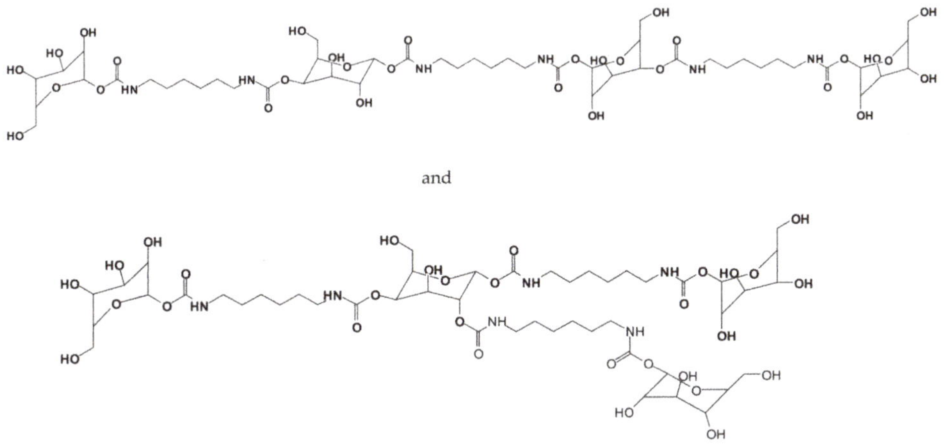

and

Figure 13. Examples of linear and branched glucose-based non-isocyanate polyurethane oligomers [121].

Glucose-based NIPUs cured at a noticeably lower temperature and were more easily spread than sucrose-based NIPUs. Wood and steel surface coatings were prepared using these NIPUs. The sessile drop test (contact angle, on wood) and the cutting grid test (on steel) of these NIPUs yielded very encouraging results [121]. Glucose NIPUs gave good results as a surface coating curing at 103 °C, whereas sucrose NIPUs performed well only at a much higher curing temperature. NIPU resins derived from saccharides have also been used as wood panel adhesives. The glucose-based NIPU gave very encouraging results for this application [121] while the sucrose-based NIPU gave very encouraging result for the preparation of polyurethane foams.

2.15. Isocyanate-Based Polyurethanes

To obtain synthetic polyurethanes using tannin as a polyol, thus by reaction with isocyanates, a first attempt was made quite early-on by direct reaction of an isocyanate with the hydroxyl groups of a condensed tannin [62]. A much more recent second approach was based on the reaction of a tannin modified either by benzoylation (Figure 14) or by oxypropylation to introduce hydroxygroups on the tannin that are more easily approachable for reaction and more likely to react as a polyol with polymeric isocyanates [63–65].

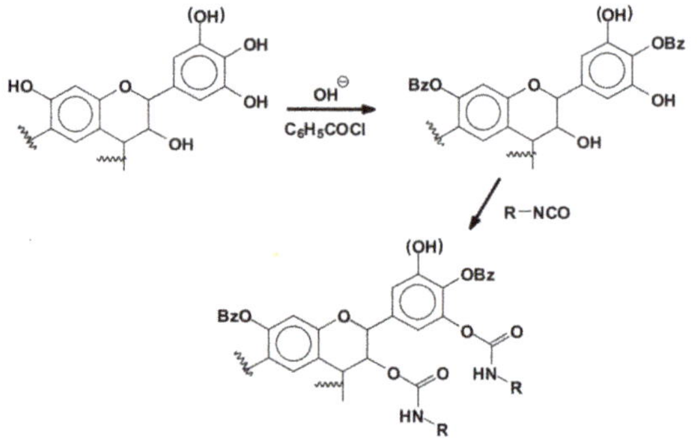

Figure 14. Reaction of a flavonoid tannin modified by benzoylation with an isocyanate.

This first approach follows the same approach that was made with another natural polyphenol, lignin [66]. This is a traditional approach where the tannin only functions as a polyol. Nevertheless, while tannin-based polyurethanes are indeed obtained by this route, other natural polyols do exist that are much more adapted to the formation of polyurethanes by reaction with isocyanates, and moreover, without needing to add an additional reaction such as oxypropylation or others.

2.16. Hard Thermoset Plastics and Resins for Angle Grinder Disks and for Automotive Brake Pads

New 100% bio-sourced, tannin/furan hard thermosetting plastics were prepared by copolymerization and characterized [122]. This new material is synthesized by the reaction of tannin and furfuryl alcohol, two inexpensive vegetable chemicals. The co-polymerization processes of both were investigated by ^{13}C NMR and matrix assisted laser desorption ionization time of flight (MALDI-TOF) mass spectrometry. The tannin/furan thermosetting resin with 100% renewable organic material has a glass transition temperature of up to 211 °C and a 95% weight loss temperature of 244 °C and 240 °C in a nitrogen atmosphere and in the air, respectively [122]. The yield of carbonization achieved is 52%. In addition, this new thermoset material has excellent mechanical properties: the Brinell hardness is superior to commercial acrylic plastics, polyvinyl chloride, and slightly lower than solid polystyrene. Compression strength was found to be as high as 194.4 MPa, thus higher than that of filled phenolic resins, and much higher than that of solid polystyrene and acetal resins [122].

A tannin–furfuryl alcohol thermoset resin [122,123] has also been used as a resin matrix for solid grinding wheels [123], these being also tested and their performance characterized and showing good abrasion comparable to commercial grinding wheels. This resin uses only bio-sourced raw materials. Moreover, its process of preparation is simple, and thus easily transferred under industrial conditions. This resin has been used to bond different mineral and organic abrasive powders to prepare the grinding wheels.

The same bio-sourced tannin–furanic thermosetting resin has been used as a resin matrix for angle grinder disks, presenting excellent abrasion and cutting properties [123] (Figure 15). A fiberglass mesh support was used for the abrasive powder and the green resin matrix yielded a rather easy manufacturing process. Aluminum dioxide was the abrasive used. The mechanical strength of these grinder disks was comparable to their commercial equivalents using synthetic phenol–formaldehyde resins. They stood up well at 11,000 rpm to the high stresses induced by cutting or grinding steel [123]. Steel bar cutting times with these green experimental disks compared well to those of commercial synthetic resin-bonded disks.

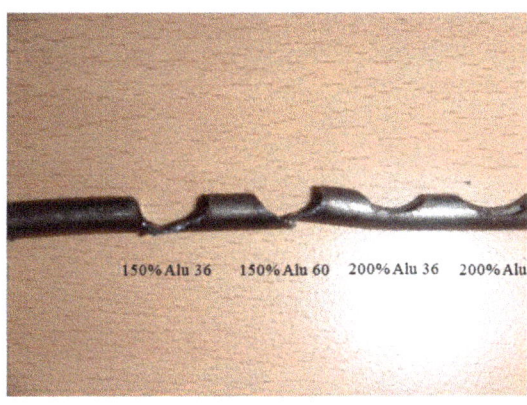

Figure 15. Example of an angle grinder disk using a tannin–furfuryl alcohol resin as a matrix, and example of cuts in a steel bar with it when using different abrasives.

Automotive brake pads bonded with a bio-sourced thermoset tannin–furanic matrix were developed and tested [124] using similar matrix resins. Their preparation is also quite easy. The braking and characteristics of such green brake pads tested under real-world large scale test conditions were excellent as well as their wear resistance [124,125]. Their performance compared well with that of synthetic phenolic resin-bonded commercial automotive brake pads. They stood up well to heavy braking, and strong stresses such as emergency braking at 50 km/h (31 mph) to a complete stop. The braking distances were comparable and sometime shorter than commercial brake pads.

2.17. Epoxy and Epoxy–Acrylic Resins

Hydrolyzed and condensed tannins have also been used to synthesize monomer epoxies. Thus, either epichlorhydrin was reacted with catechin to yield an epoxidized catechin monomer or alternatively by alkylating it with unsaturated halogenated compounds and then oxidizing [126]. Analysis of the reaction products shows the presence of a by-product of the benzodioxane group, which then decreases the average epoxy functionality [127].

Following the excellent seminal works cited above, more recent work has appeared on the preparation and characterization of condensed tannin epoxies [128,129]. One of these deals with the preparation an epoxy–acrylic tannin resin capable of quick curing without using a hardener [128,129] (Figure 16). Glycidyl ether tannin (GET) was the starting material for preparing the tannin epoxy acrylate resin by reacting it with acrylic acid with a catalyst and adding hydroquinone as well [129] (Figure 16). It was tested for shear resistance with very interesting results. Unlike the other studies that used monomers as models, these two studies were conducted with a commercial extract of mimosa tannin, thus in a real situation.

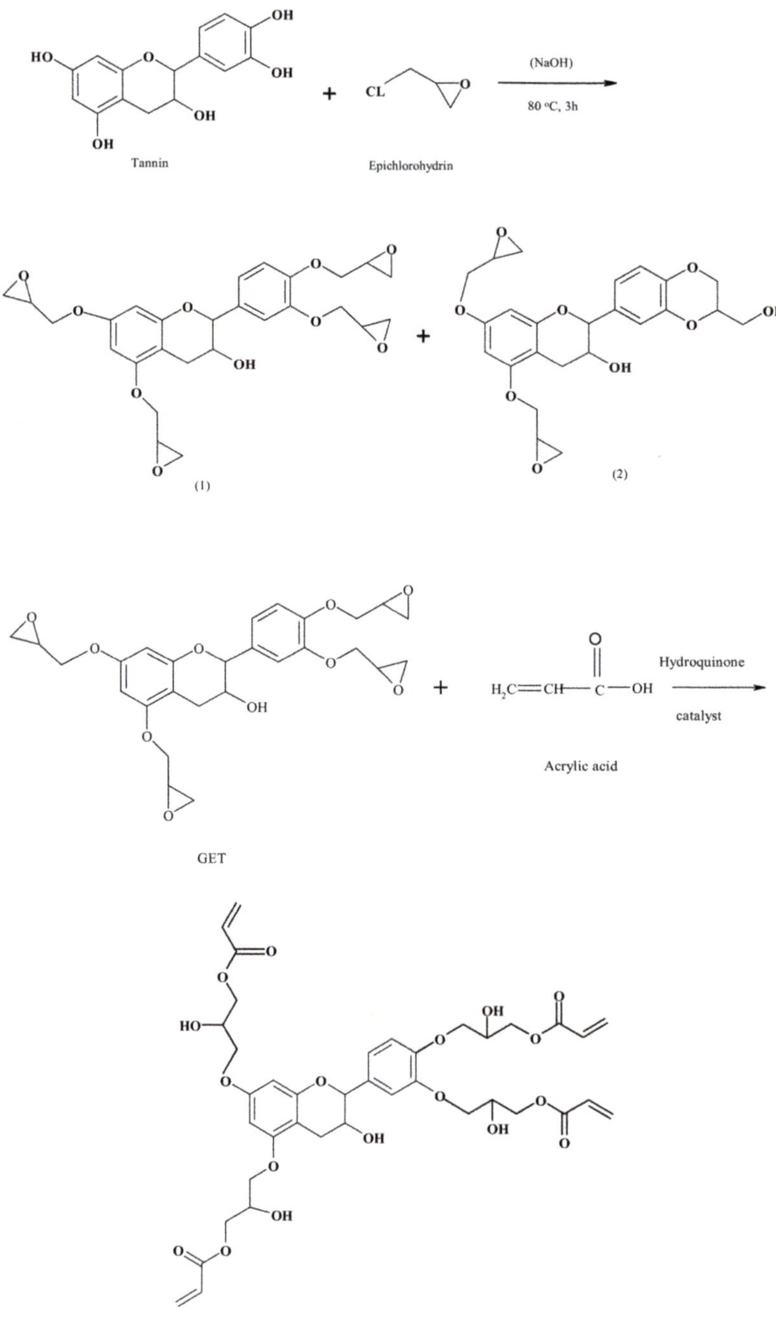

Figure 16. Scheme of the sequence of reactions to prepare the epoxy acrylate resin. The reaction proceeds in two steps: first, the epoxidization of the flavonoid tannin, and second, its reaction with acrylic acid catalyzed by hydroquinone [129].

Tomita and Yanezawa first reported the epoxidation of gallic acid with epichlorohydrin [130]. The presence of an ammonium catalyst of phase transfer under anhydrous conditions promotes the addition of epichlorhydrin to the carboxylic acid function and to one phenol group at least. The epoxy equivalent weights (EEW) reported were between 137 and 160. This corresponds to a 1 to 4 epoxy functionality. The epoxidized compound is then cross-linked with a conventional crosslinking agent polyamine, or an anhydride is then used to cross-link the epoxy resin formed, thus by a traditional crosslinker [130]. Theoretical calculations, however, did show that Tomita and Yanezawa achieved an average epoxy functionality of 2, notwithstanding the use of four epichlorhydrins per phenol group.

The synthesis conditions of gallic tetra epoxy acid has been recently published [131]. Only resins with only two or three epoxy groups were reported before. This work established a relationship between different phenol monomer chemical structures and their reactivity with epichlorhydrin by studying the mechanism of O-glycidylation. Molecular understanding of this relationship is determinant for developing bio-sourced epoxy monomers derived from tannins [131].

Nouailhas et al. [127] have also recently proposed another synthetic route leading to epoxy gallic acid allylation prepolymers by reacting gallic acid with allyl bromide followed by double bond oxidation using m-chloro-perbenzoic acid. This method makes it possible to obtain epoxy prepolymers with epoxy functionality up to 3.

Green tea condensed tannins have also been proposed as suitable to synthesize bio-sourced aromatic epoxy oligomers, based on work on model molecules of tannins [132].

2.18. Tannin-Impregnated Fibreboards

Nonwoven flax and hemp fiber mats were impregnated with renewable bio-sourced resin matrices to yield both high and low density composites of good performance [133–136] (Figure 17). The types of bio-sourced matrices used were: (1) a 5% hexamethylenetetramine-hardened resin based on commercial mimosa tannin extract, and (2) a 50/50 combination by weight of glyoxalated organosolv lignin of low molecular weight and of mimosa tannin with hexamethylenetetramine. Modulus of Elasticity (MOE) in bending and tensile strength were tested for these composites to maximize tensile strength. Corona-treatment was also used to improve the tensile strength of such composites, and its most adequate duration determined. These composites were tested for different characteristics, such as surface hardness, water contact angles, influence of fiber morphology, and others, always yielding good results. The matrices based on mixed tannin and lignin gave composites with a thermoplastic behavior after just the first hotpressing. They were thus thermoformable as they could be shaped in their final form by a second hotpressing.

Figure 17. A nonwoven web of hemp fibers and the composite obtained by impregnation with a tannin–furanic resin.

The hardening cycle, press temperature, press time, pressure, moisture content, and the number of fiber mats influenced the composite characteristics. Furthermore, first drying the resin-impregnated fiber mats for storing them and then rehydrating them before hotpressing allowed the mats to be maintained ready for use even after long storage times. This approach still yielded composites of 50% matrix resin/50% natural fibers presenting good tensile strength, water swelling, and Young's modulus. The best results were achieved at a slow curing low temperature (130 °C, 35 min) at a 20% moisture content.

Acid catalyzed tannin/furfuryl alcohol resins in proportion 45/54% by weight were also tried for this application, yielding flax fiber lightweight composites with good performance.

2.19. Wood Surface Finishes

Tannin–furanic resin impregnated paper was applied as a surface finish to plywood with good results. Hexamethylenetetramine, paraformaldehyde, formurea, and a mix of the latter and paraformaldehyde were used as hardeners. These gave good results in both standard crosscut and water vapor tests. Chemical analysis has confirmed both the reaction of tannin with furfuryl alcohol and the autocondensation of the latter, as well as the co-reaction of both tannin and furanic aromatic rings with formaldehyde-yielding compounds such as hexamine. The oligomers formed were determined [137].

MUF impregnated paper surface coatings were compared with the tannin–furanic impregnated papers with regards to the surface quality of wood panels, with the latter giving better water vapor resistance and cross-cut testing. Abrasion resistance was influenced for both types of coatings by the impregnated papers moisture content. Color measurements indicated that a higher moisture content yielded a lighter color of mimosa tannin impregnated papers. Nonetheless, impregnated surfaces with tannin/furan resin impregnated paper are too dark for decorative applications, but may be useful for the manufacture of cement formwork panels [138].

By paper impregnation with a tannin–furanic resin hardened with a formurea concentrate, high pressure paper laminates were prepared. These were tested for crosscut, abrasion, and water vapor resistance. Ten-layer high pressure paper laminates of this type improved the dry shear strength of plywood to which they were applied as a surface coating as well reducing its water absorption. Paper laminates pressed for 600 s, at a 140 °C and 120 kg/cm^2, yielded the best results. High pressure laminates impregnated with a tannin/furanic resin and creating thick materials compare favorably with the properties of synthetic high pressure thick laminates impregnated with synthetic phenol–formaldehyde resins as used in mechanical gears, etc. [139].

2.20. Flexible Plastic Films

Flexible films and strongly adherent surface finishes have been prepared [140] by reacting partially aminated polyfluorocarbon tannins with furfuryl alcohol in the presence of plasticizers such as glycerol or polyethyleneimine. 13C NMR analysis shows a partial amination of the tannin under the conditions used and even the formation of –N= bridges between the flavonoids, although these have proved to be rare [102]. Thus structures as in Figure 18 have been observed [141].

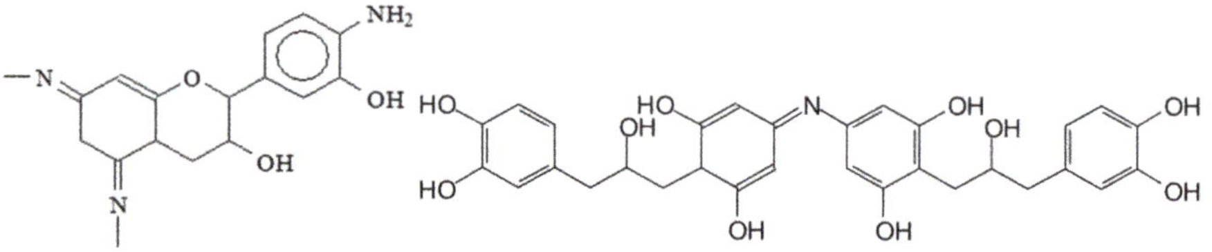

Figure 18. Cont.

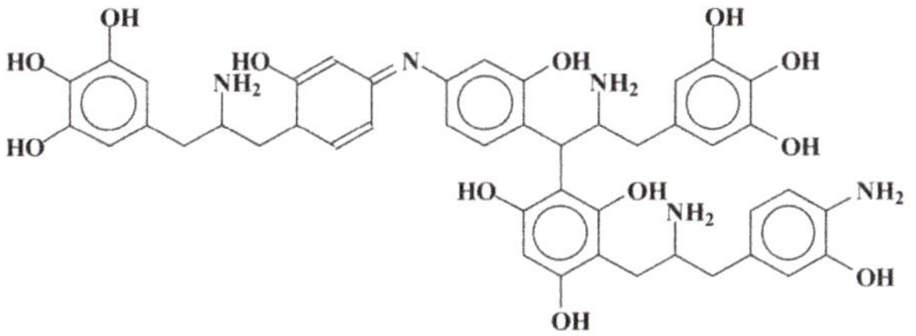

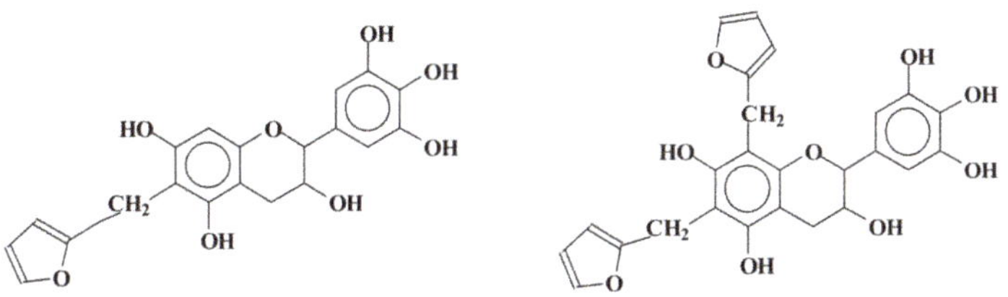

Figure 18. Examples of compounds prepared by amination of a flavonoid tannin, and simultaneous reaction of the tannin with furfuryl alcohol.

MALDI-TOF analysis has shown the presence of oligomers produced by the reaction of furfuryl alcohol with flavonoid A-rings and the simultaneous self-condensation of furfuryl alcohol [140]. Thus, the methylene–furanic linear chains have also been shown to be linked to reactive flavonoid tannin sites. In addition, side condensation reactions of furfuryl alcohol led to the formation of methylene ether bridges between furanic rings, followed by rearrangement to methylene bridges with consequent release of formaldehyde. The latter reacted with both the flavonoid and furan reactive sites to give –CH$_2$OH and –CH$_2$+ groups and further methylene bridges [140].

2.21. Cement Superplasticizers

To avoid the retardation effect that common plasticizers have while fludifying cement, superplasticizers have been developed so that this retardation effect could be eliminated. They improve the facility of working cement, reduce the water needed, and improve the cement final strength. These materials transform cement pastes into flowing fluids. Commercial superplasticizers are synthetic resins such as sulfonated melamine–formaldehyde or naphthalene sulfonate–formaldehyde [52–54]. Modified lignosulphonates are also used for this category of materials [142–144]. Up to 30% water reduction are possible with slump sizes of 200 mm.

Their mechanism is based on their adsorption on the cement grain surface while maintaining the water orientation of their sulphonic groups. The water monolayer that is formed around the grain causes a dispersion of the grains contributing to the fluidification of the cement/concrete paste. Moreover, the surface tension of water is not much reduced and there is no significant retardation of cement setting or hardening [144].

The structures of polyflavonoid tannins have the ability to complex Fe^{2+}/Fe^{3+} and aluminum ions in cement by the ortho-diphenol hydroxyls on their B-rings [2,10]. Sulphonation improves their solubility in water [2,10]. These characteristics make polyflavonoid tannins attractive as dispersants/superplasticizers for cement, composed mainly of Ca and Fe silicates and aluminates.

Several tannins, especially sulphited, have performed well as cement superplasticizers [145]. Extracts of sulphited mimosa, quebracho, and pine tannin all behave very well as cement superplasticizers, with mimosa and pine being those with slightly better behavior. A dosage of 0.25 to 0.5% by weight on the cement has a significant effect of fluidification.

Modified condensed tannin extracts behave as superplasticizers. Cement and concrete flow better and the onset of hardening is not retarded [145]. The total effect was due to a mix of different causes: (1) the silicate and aluminate cement components inducing an increase in the tannin molecular weight,

(2) the improvement in tannin solubility due to the insertion of sulfonic groups leading first to the decrease in the tannin's molecular mass and then to its stabilization, and (3) urea addition stabilizing the molecular mass by minimizing tannin colloidal interactions and hindering tannin rearrangement and autocondensation [145].

2.22. Ferric Tannate Inks

Ferric gallo-tannate ink is a black to purple ink made from metal salts, especially ferrous sulfate but sometimes copper sulfate, and various tannins of vegetable origin. Black ink, emblematic of the monastic scriptorium, was the most used ink in Europe between the twelfth and nineteenth centuries. This tannic ink or solubilized tannins is sometimes referred to as ferric ink, ferro-gallic, or metallogallic. Irreversible damage to paper due to this corrosive ink poses significant conservation problems. The particularity of this ink lies in its absence of pigment or dye; it is the action of metal salts (iron or copper sulfate) which, added to the tannic material (the gall nut), gives the black hue, most often dark violet before aging. Its defect is its corrosivity, for the paper as well as for the metal quill. To limit this inconvenience, its manufacture must be the subject of a good aging; maceration of the tannin for three months then, after mixing with the iron salt, a maturation of at least two months to one year guaranteeing its optimum. There is a plethora of different recipes. The three main constituents are:

(1) the gall nut (gallo-tannic acid) or various solubilized extracts of bark tannins (oak, etc.) which are possibly dried, (2) iron sulfate (ferrous sulfate). Sometimes copper sulfate can be used, but this, also corrosive, attacks the paper, and (3) a binder, such as gum arabic or wine lees (that is to say, wine whose liquid part has evaporated). This mixture is swelled in lukewarm water for one day. Salicylic acid or phenol may be added to prevent the development of microorganisms. It is hygroscopic (slows down drying) and keeps the ink particles in suspension and prevents them from precipitating (colloidal solution). In the fifteenth century, plum, apricot, or cherry resins were used to achieve this. The excess of ink so prepared can be dried. Full drying was done in pork bladders. The base ink (tannin) is then redissolved and a reagent is prepared, with iron sulfate dissolved in warm water, to which gum arabic is added. The tannin solution is boiled separately and gum arabic is also added. The mixture is allowed to cool and the two solutions are mixed. The ink so prepared is then ready for writing. This type of ink was still used in primary schools in some countries in Europe for some time after the Second World War.

3. Conclusions

The many uses and application of tannins described above need to be put into perspective with regards to possible further advances, existing drawbacks, and future potential. Of all the applications described above, leather tanning is still the main industrial use of vegetable tannins. While their use for this application has been progressively decreasing and limited to heavy duty leathers, as displaced for finer applications by chrome tanning, the real or perceived toxicity of chrome has spurred considerable research on feasible alternatives. Not all of these have gone in the direction of vegetable tanning, as synthetic resins such as melamine-based ones have been considered. However, vegetable tanning has regained some interest for soft leathers either in combination with oils or with synthetic resins. Although further development in this application for vegetable tannins cannot be considered as static, nonetheless the potential for future expansion is rather limited.

The second most important application of vegetable tannins is in wood adhesives. The strong shift away from synthetic resins based on formaldehyde has favored the interest in the use of tannins as well as of other natural raw materials for this application. The tannin adhesive technology is definitely more advanced, and more used, than other bio-sourced materials, having proved itself industrially in several countries over a number of decades. Its drawback at present is that their present supply is limited. The potential world supply of tannins is really huge, what is lacking, however, is a marked increase in the factories extracting them. A few new tannin extraction factories have been created in the last decade, but competition with other bio-sourced materials already industrially available either as waste or obtained by other already existing production sources is rather intense at present.

Medical and pharmaceutical applications are one of the more interesting and active fields of research at present for the evaluation of vegetable tannins. While some pharmaceutical applications already exist, for further progress the results being developed in this field need to be proven in vivo, this being an important phase of development. It is difficult to say for what specific pharmaceutical uses tannin might be successfully adopted. The main drawback here is that the balance of properties favorable and unfavorable to each application have to be evaluated. It is nonetheless an application that is likely to further flourish in the future.

More established and fully functional is the use of tannins in the beverage industry, be it wine, beer, or fruit juices. Their use will expand with the expansion of these markets due to the expansion of the population, but not for different applications in the field.

Tannin based foams, be it the more developed phenolic–furanic type, isocyanate-derived polyurethanes, or the newer, less developed non-isocyanate poly(hydroxy)urethanes, is a fast moving research field for thermal and/or acoustic insulation, for hydroponics, and a number of other applications. A considerable amount of research is still going on in this field, and some industrial trials too, but all this has not as yet materialized in an industrial application.

Tannin-based antipollution flocculants and corrosion inhibitors have been developed quite a long time ago, in the late 1960's and early 1970's and used industrially for some time. After a period of having practically disappeared from the market they are regaining favor, both in research and industrially, due to the interest in substituting bio-sourced material for somewhat toxic or oil derived synthetic materials.

As regards the other applications, foundry sand binders is used but it is unlikely to increase in market share due to the competition of other more performant materials. The same is valid for drilling fluids. Corrugated cardboard adhesives are used in a few developing countries for a niche industrial market, namely the moisture proofing of starch-bonded corrugated cardboard boxes for fruit exports having to pass through humid conditions, such as in the tropics.

Very new is the development of adhesives to bind teflon to steel and aluminum. While the technology exists and has proven itself it seems that the moment for a bio-sourced adhesive for this application has not yet come, although patents on the subject have been created. The writer supposes that this technology might eventually be used industrially once environmental protection awareness becomes stronger, and stricter environmental protection rules may force its application.

The situation is the same for the hard plastics used as matrices for abrasive angle grinders, discs, and automotive brake pads. Only time will tell if these developments, some of them patented, will ever reach industrial use.

Epoxy resins based on tannins have been developed by a few groups, one of which is in direct contact with interested industries. It is likely that some industrial development will eventually arise from this line of research, although none is known to date.

All the other applications are all in the purely experimental phase, and it is difficult to see if they will ever develop further or not.

Finally, ferric inks, the main source of writing inks for several past centuries, is definitely out of interest as more performant materials exist today, and no further interest in them is apparent.

Funding: This research received no external funding.

Conflicts of Interest: The authors declare no conflict of interest.

References

1. Caller, L. *Le Fabbriche Italiane di Estratto di Castagno*; Silva Chimica: San Michele Mondovi (CN), Italy, 1989.
2. Pizzi, A. Tannin-based wood adhesives. In *Wood Adhesives Chemistry and Technology*; Pizzi, A., Ed.; Marcel Dekker: New York, NY, USA, 1983; Volume 1, pp. 178–246.

3. Giovando, S.; Pizzi, A.; Pasch, H.; Pretorius, N. Structure and oligomers distribution of commercial Tara (*Caesalpina spinosa*) hydrolysable tannin. *ProLigno* **2013**, *9*, 22–31.
4. Radebe, N.; Rode, K.; Pizzi, A.; Giovando, S.; Pasch, H. MALDI-TOF-CID for the Microstructure Elucidation of Polymeric Hydrolysable Tannins. *J. Appl. Polym. Sci.* **2013**, *128*, 97–107. [CrossRef]
5. Pizzi, A.; Pasch, H.; Rode, K.; Giovando, S. Polymer structure of commercial hydrolisable tannins by MALDI-TOF mass spectrometry. *J. Appl. Polym. Sci.* **2009**, *113*, 3847–3859. [CrossRef]
6. Pasch, H.; Pizzi, A. On the macromolecular structure of chestnut ellagitannins by MALDI-TOF mass spectrometry. *J. Appl. Polym. Sci.* **2002**, *85*, 429–437. [CrossRef]
7. Drewes, S.E.; Roux, D.G. Condensed tannins. 15. Interrelationships of flavonoid components in wattle-bark extract. *Biochem. J.* **1963**, *87*, 167–172. [CrossRef] [PubMed]
8. Roux, D.G.; Paulus, E. Condensed tannins. 8. The isolation and distribution of interrelated heartwood components of *Schinopsis* spp. *Biochem. J.* **1961**, *78*, 785–789. [CrossRef]
9. Saayman, H.M.; Roux, D.G. The origins of tannins and flavonoids in black-wattle barks and heartwoods, and their associated 'non-tannin' components. *Biochem. J.* **1965**, *97*, 794–801. [CrossRef]
10. Pizzi, A. *Advanced Wood Adhesives Technology*; Marcel Dekker: New York, NY, USA, 1994; pp. 149–218.
11. Pizzi, A. Wattle-based adhesives for exterior grade particleboard. *Forest Prod. J.* **1978**, *28*, 42–47.
12. Pizzi, A.; Scharfetter, H. The chemistry and development of tannin-based wood adhesives for exterior plywood. *J. Appl. Polym. Sci.* **1978**, *22*, 1745–1761. [CrossRef]
13. Valenzuela, J.; von Leyser, E.; Pizzi, A.; Westermeyer, C.; Gorrini, B. Industrial production of pine tannin-bonded particleboard and MDF. *Eur. J. Wood Wood Prod.* **2012**, *70*, 735–740. [CrossRef]
14. Pizzi, A. Glue blenders effect on particleboard using wattle tannin adhesives. *Holzforschung Holzverwertung* **1979**, *31*, 85–86.
15. Pizzi, A. Hot-setting tannin-urea-formaldehyde exterior wood adhesives. *Adhes. Age* **1977**, *20*, 27–32.
16. Custers, P.A.J.L.; Rushbrook, R.; Pizzi, A.; Knauff, C.J. Industrial applications of wattle-tannin/urea-formaldehyde fortified starch adhesives for damp-proof corrugated cardboard. *Holzforschung Holzverwertung* **1979**, *31*, 131–132.
17. Pizzi, A.; Roux, D.G. The chemistry and development of tannin-based weather- and boil-proof cold-setting and fast-setting adhesives for wood. *J. Appl. Polym. Sci.* **1978**, *22*, 1945–1954. [CrossRef]
18. Pizzi, A.; Rossouw, D.D.T.; Knuffel, W.; Singmin, M. "Honeymoon" phenolic and tannin-based fast setting adhesive systems for exterior grade fingerjoints. *Holzforschung Holzverwertung* **1980**, *32*, 140–151.
19. Pizzi, A.; Cameron, F.A. Fast-set adhesives for glulam. *Forest Prod. J.* **1984**, *34*, 61–65.
20. Mansouri, H.R.; Pizzi, A.; Fredon, E. Honeymoon fast-set adhesives for glulam/fingerjoints of higher natural materials content. *Eur. J. Wood Wood Prod.* **2009**, *67*, 207–210. [CrossRef]
21. Pizzi, A. Phenolic resins by reactions of coordinated metal ligands. *J. Polym. Sci. Polym. Lett. Ed.* **1979**, *17*, 489–492. [CrossRef]
22. Pizzi, A. Phenol and tannin-based adhesive resins by reactions of coordinated metal ligands, Part 1: Phenolic chelates. *J. Appl. Polym. Sci.* **1979**, *24*, 1247–1255. [CrossRef]
23. Pizzi, A. Phenol and tannin-based adhesive resins by reactions of coordinated metal ligands, Part II: Tannin adhesives preparation, characteristics and application. *J. Appl. Polym. Sci.* **1979**, *24*, 1257–1268. [CrossRef]
24. Von Leyser, E.; Pizzi, A. The formulation and commercialization of glulam pine tannin adhesives in Chile. *Holz als Roh-und Werkstoff* **1990**, *48*, 25–29. [CrossRef]
25. Pizzi, A.; Walton, T. Non-emulsifiable, water-based diisocyanate adhesives for exterior plywood, Part 1: Novel reaction mechanisms and their chemical evidence. *Holzforschung* **1992**, *46*, 541–547. [CrossRef]
26. Pizzi, A.; Valenzuela, J.; Westermeyer, C. Non-emulsifiables, water-based, diisocyanate adhesives for exterior plywood, Part 2: Industrial application. *Holzforschung* **1993**, *47*, 69–72. [CrossRef]
27. Böhm, R.; Hauptmann, M.; Pizzi, A.; Friederich, C.; Laborie, M.-P. The chemical, kinetic and mechanical characterization of Tannin-based adhesives with different crosslinking systems. *Int. J. Adhes. Adhes.* **2016**, *68*, 1–8. [CrossRef]
28. Santiago-Medina, F.J.; Foyer, G.; Pizzi, A.; Calliol, S.; Delmotte, L. lignin-derived non-toxic aldehydes for ecofriendly tannin adhesives for wood panels. *Int. J. Adhes. Adhes.* **2016**, *70*, 239–248. [CrossRef]
29. Pizzi, A.; Meikleham, N.; Dombo, B.; Roll, W. Autocondensation-based, zero-emission, tannin adhesives for particleboard. *Holz als Roh-und Werkstoff* **1995**, *53*, 201–204. [CrossRef]

30. Meikleham, N.; Pizzi, A.; Stephanou, A. Induced accelerated autocondensation of polyflavonoid tannins for phenolic polycondensates, Part 1: 13C NMR, 29Si NMR, X-ray and polarimetry studies and mechanism. *J. Appl. Polym. Sci.* **1994**, *54*, 1827–1845. [CrossRef]
31. Pizzi, A.; Meikleham, N.; Stephanou, N. Induced accelerated autocondensation of polyflavonoid tannins for phenolic polycondensates-Part II: Cellulose effect and application. *J. Appl. Polym. Sci.* **1995**, *55*, 929–933. [CrossRef]
32. Garcia, R.; Pizzi, A.; Merlin, A. Ionic polycondensation effects on the radical autocondensation of polyflavonoid tannins-An ESR study. *J. Appl. Polym. Sci.* **1997**, *65*, 2623–2632. [CrossRef]
33. Garcia, R.; Pizzi, A. Polycondensation and autocondensation networks in polyflavonoid tannins, Part 1: Final networks. *J. Appl. Polym. Sci.* **1998**, *70*, 1083–1091. [CrossRef]
34. Garcia, R.; Pizzi, A. Polycondensation and autocondensation networks in polyflavonoid tannins, Part 2: Polycondensation vs. autocondensation. *J. Appl. Polym. Sci.* **1998**, *70*, 1093–1110. [CrossRef]
35. Garcia, R.; Pizzi, A. Cross-linked and entanglement networks in thermomechanical analysis of polycondensation resins. *J. Appl. Polym. Sci.* **1998**, *70*, 1111–1116. [CrossRef]
36. Pichelin, F.; Nakatani, M.; Pizzi, A.; Wieland, S.; Despres, A.; Rigolet, S. Structural beams from thick wood panels bonded industrially with formaldehyde free tannin adhesives. *Forest Prod. J.* **2006**, *56*, 31–36.
37. Navarrete, P.; Mansouri, H.R.; Pizzi, A.; Tapin-Lingua, S.; Benjelloun-Mlayah, B.; Rigolet, S. Synthetic-resin-free wood panel adhesives from low molecular mass lignin and tannin. *J. Adhes. Sci. Technol.* **2010**, *24*, 1597–1610. [CrossRef]
38. Ghahri, S.; Pizzi, A.; Mohebby, B.; Mirshoktaie, A.; Mansouri, H.R. Soy-based, tannin-modified plywood adhesives. *J. Adhes.* **2018**, *94*, 218–237. [CrossRef]
39. Abdullah, U.H.B.; Pizzi, A. Tannin-Furfuryl alcohol wood panel adhesives without formaldehyde. *Eur. J. Wood Wood Prod.* **2013**, *71*, 131–132. [CrossRef]
40. Ballerini, A.; Despres, A.; Pizzi, A. Non-toxic, zero-emission tannin-glyoxal adhesives for wood panel. *Holz als Roh-und Werkstoff* **2005**, *63*, 477–478. [CrossRef]
41. Trosa, A.; Pizzi, A. A no-aldehyde emission hardener for tannin-based wood adhesives. *Holz als Roh-und Werkstoff* **2001**, *59*, 266–271. [CrossRef]
42. Grigsby, W.J.; McIntosh, C.D.; Warnes, J.M.; Suckling, I.D.; Anderson, C.R. Adhesives. U.S. Patent 7,319,115 B2, 15 January 2008.
43. Trosa, A.; Pizzi, A. Industrial hardboard and other panels binder from tannin/furfuryl alcohol in absence of formaldehyde. *Holz als Roh-und Werkstoff* **1998**, *56*, 213–214. [CrossRef]
44. Kamoun, C.; Pizzi, A. Mechanism of hexamine as a non-aldehyde polycondensation hardener, Part 1: Hexamine decomposition and reactive intermediates. *Holzforschung Holzverwertung* **2000**, *52*, 16–19.
45. Kamoun, C.; Pizzi, A. Mechanism of hexamine as a non-aldehyde polycondensation hardener, Part 2: Recomposition of intermediate reactive compound. *Holzforschung Holzverwertung* **2000**, *52*, 66–67.
46. Kamoun, C.; Pizzi, A.; Zanetti, M. Upgrading of MUF resins by buffering additives—Part 1: Hexamine sulphate effect and its limits. *J. Appl. Polym. Sci.* **2003**, *90*, 203–214. [CrossRef]
47. Yang, L.L.; Wang, C.C.; Yea, K.-Y.; Yoshida, T.; Hatano, T.; Okada, T. Antitumor activity of ellagitannins on tumor cell lines. In *Plant Polyphenols 2*; Gross, G.G., Hemingway, R.W., Yoshida, T., Eds.; Kluwer Academic/Plenum Publishers: New York, NY, USA, 1999; pp. 615–628.
48. Nakamura, Y.; Matsuda, M.; Honma, T.; Tomita, I.; Shibata, N.; Warashina, N.; Noro, T.; Hara, Y. Chemical constituents of mainly active components fractionated from the aqueous tea non-dialysates, an antitumor promoter. In *Plant Polyphenols 2*; Gross, G.G., Hemingway, R.W., Yoshida, T., Eds.; Kluwer Academic/Plenum Publishers: New York, NY, USA, 1999; pp. 629–642.
49. Miyamoto, K.; Murayama, T.; Hatano, T.; Yoshida, T.; Okuda, T. Host-mediated anticancer activity of tannins. In *Plant Polyphenols 2*; Gross, G.G., Hemingway, R.W., Yoshida, T., Eds.; Kluwer Academic/Plenum Publishers: New York, NY, USA, 1999; pp. 643–644.
50. Noro, T.; Ohki, T.; Noda, Y.; Warashina, T.; Noro, K.; Tomita, I.; Nakamura, Y. Inhibitory effect of hydrolysable tannins on tumor promoting activities of 12-O-tetradecanoylphorbol-13-acetate (TPA) in JB6 mouse epidermal cells. In *Plant Polyphenols 2*; Gross, G.G., Hemingway, R.W., Yoshida, T., Eds.; Kluwer Academic/Plenum Publishers: New York, NY, USA, 1999; pp. 665–674.
51. Kashiwada, Y.; Nonaka, G.; Nishioka, I.; Chang, J.J.; Lee, K.H. Antitumor agents, 129. Tannins and related compounds as selective cytotoxic agents. *J. Nat. Prod.* **1992**, *55*, 1033–1043. [CrossRef] [PubMed]

52. Pizzi, A. Tannins: Major Sources, Properties and Applications. In *Monomers, Polymers and Composites from Renewable Resources*; Belgacem, M.N., Gandini, A., Eds.; Elsevier: Oxford, UK, 2008; pp. 179–199.
53. Krifa, M.; Pizzi, A.; Mousli, M.; Chekir-Ghedira, L.; Leloup, L.; Ghedira, K. *Limoniastrum guyonianum* aqueous gall extract induces apoptosis in colorectal cancer cells by inhibiting calpain activity. *Tumor Biol.* **2014**, *35*, 7877–7885. [CrossRef] [PubMed]
54. Quideau, S.; Jourdes, M.; Saucier, C.; Glories, Y.; Pardon, P.; Baudry, C. DNA topoisomérase inhibitor acutissimin A and other flavano-ellagitannins in red wine. *Angew. Chem.* **2003**, *42*, 6012–6014. [CrossRef]
55. Quideau, S.; Jourdes, M.; Lefeuvre, D.; Montaudon, D.; Saucier, C.; Glories, Y.; Pardon, P.; Pourquier, D. The Chemistry of Wine Polyphenolic C-Glycosidic Ellagitannins Targeting Human Topoisomerase II. *Chem. Eur. J.* **2005**, *11*, 6503–6513. [CrossRef] [PubMed]
56. Funatogawa, K.; Hayashi, S.; Shimomura, H.; Yoshida, T.; Hatano, T.; Ito, H.; Hirai, Y. Antibacterial activity of hydrolyzable tannins derived from medicinal plants against Helicobacter pylori. *Microbiol. Immunol.* **2004**, *48*, 251–261. [CrossRef]
57. Fiori, G.M.L.; Fachin, A.L.; Correa, V.S.C.; Bertoni, B.W.; Giuliatti, S.; Amui, S.F.; Franca, S.d.C.; Pereira, A.M.S. Antimicrobial Activity and Rates of Tannins in Stryphnodendron adstringens Mart. Accessions Collected in the Brazilian Cerrado. *Am. J. Plant Sci.* **2013**, *4*, 2193–2198. [CrossRef]
58. Audi, E.A.; Toledo, D.P.; Peres, P.G.; Kimura, E.; Pereira, W.K.V.; de Mello, J.C.P.; Nakamura, C.; Alves-do-Prado, W.; Cuman, R.K.N.; Bersani, C.A. Gastric antiulcerogenic effects of Stryphnodendron adstringens in rats. *Phytother. Res.* **1999**, *13*, 264–266. [CrossRef]
59. Lai, J.C.Y.; Lai, H.Y.; Nalamolu, K.R.; Ng, S.F. Treatment for diabetic ulcer wounds using a fern tannin optimized hydrogel formulation with antibacterial and antioxidative properties. *J. Ethnopharmacol.* **2016**, *189*, 277–289. [CrossRef]
60. Saito, M.; Hosoyama, H.; Ariga, T.; Kataoka, S.; Yamaji, N. Antiulcer Activity of Grape Seed Extract and Procyanidins. *J. Agric. Food Chem.* **1998**, *46*, 1460–1464. [CrossRef]
61. Ricci, A.; Parpinello, G.; Schwertner, A.P.; Teslic, N.; Brilli, C.; Pizzi, A.; Versari, A. Quality assessment of food grade plant extracts using MALDI-TOF-MS, ICP-MS and spectrophotometric methods. *J. Food Compos. Anal.* **2017**, *59*, 95–104. [CrossRef]
62. Pizzi, A. Tannin-based polyurethane adhesives. *J. Appl. Polym. Sci.* **1979**, *23*, 1889–1990. [CrossRef]
63. Garcia, D.; Glasser, W.; Pizzi, A.; Osorio-Madrazo, A.; Laborie, M.-P. Synthesis and physicochemical properties of hydroxypropyltannins from maritime pine bark (*Pinus pinaster* Ait.). *Holzforschung* **2014**, *68*, 411–418. [CrossRef]
64. Garcia, D.; Glasser, W.; Pizzi, A.; Paczkowski, S.; Laborie, M.-P. Substitution pattern elucidation of hydroxypropyl Pinus pinaster (Ait.) bark polyflavonoids derivatives by ESI(-)-MS/MS. *J. Mass Spectrom.* **2014**, *49*, 1050–1058. [CrossRef] [PubMed]
65. Garcia, D.; Glasser, W.; Pizzi, A.; Paczkowski, S.; Laborie, M.-P. Hydroxypropyl tannin from Pinus pinaster bark as polyol source in urethane chemistry. *Eur. Polym. J.* **2015**, *67*, 152–165. [CrossRef]
66. Wu, L.; Glasser, W. Engineering plastics from lignin. I. Synthesis of hydroxypropyl lignin. *J. Appl. Polym. Sci.* **1984**, *29*, 1111–1123. [CrossRef]
67. Pizzi, A.; von Leyser, E.P.; Valenzuela, J.; Clark, J.G. The chemistry and development of pine tannin adhesives for exterior particleboard. *Holzforschung* **1993**, *47*, 164–172. [CrossRef]
68. Basso, M.C.; Pizzi, A.; Lacoste, C.; Delmotte, L.; Al-Marzouki, F.A.; Abdalla, S.; Celzard, A. Tannin-furanic-polyurethane foams for industrial continuous plant lines. *Polymers* **2014**, *6*, 2985–3004. [CrossRef]
69. Meikleham, N.; Pizzi, A. Acid and alkali-setting tannin-based rigid foams. *J. Appl. Polym. Sci.* **1994**, *53*, 1547–1556. [CrossRef]
70. Basso, M.C.; Li, X.; Giovando, S.; Fierro, V.; Pizzi, A.; Celzard, A. Green, formaldehyde-free, foams for thermal insulation. *Adv. Mat. Lett.* **2011**, *2*, 378–382. [CrossRef]
71. Li, X.; Pizzi, A.; Zhou, X.; Fierro, V.; Celzard, A. Formaldehyde-free prorobitenidin/profisetinidin tannin/furanic foams based on alternative aldehydes: Glyoxal and glutaraldehyde. *J. Renew. Mater.* **2015**, *3*, 142–150. [CrossRef]
72. Basso, M.C.; Giovando, S.; Pizzi, A.; Celzard, A.; Fierro, V. Tannin/furanic foams without blowing agents and formaldehyde. *Ind. Crops Prods.* **2013**, *49*, 17–22. [CrossRef]
73. Basso, M.C.; Giovando, S.; Pizzi, A.; Lagel, M.C.; Celzard, A. Alkaline tannin rigid foams. *J. Renew. Mater.* **2014**, *2*, 182–185. [CrossRef]

74. Lagel, M.C.; Pizzi, A.; Giovando, S.; Celzard, A. Development and characterization of phenolic foams with phenol-formaldehyde-chestnut tannins resin. *J. Renew. Mater.* **2014**, *2*, 220–229. [CrossRef]
75. Li, X.; Basso, M.C.; Fierro, V.; Celzard, A. Chemical modification of tannin/furanic rigid foams by isocyanates and polyurethanes. *Maderas Cienc. Technol.* **2012**, *14*, 257–265. [CrossRef]
76. Lacoste, C.; Basso, M.C.; Pizzi, A.; Laborie, M.-P.; Celzard, A. Natural albumin/tannins cellular foams. *Ind. Crops Prods.* **2015**, *73*, 41–48. [CrossRef]
77. Basso, M.C.; Lagel, M.C.; Pizzi, A.; Celzard, A.; Abdalla, S. First tools for tannin-furanic foams design. *Bioresources* **2015**, *10*, 5233–5241. [CrossRef]
78. Basso, M.C.; Pizzi, A.; Celzard, A. Influence of formulation on the dynamics of preparation of tannin based foams. *Ind. Crops Prod.* **2013**, *51*, 396–400. [CrossRef]
79. Rangel, G.; Chapuis, H.; Basso, M.C.; Pizzi, A.; Delgado, C.; Fierro, V.; Celzard, A.; Gerardin, C. Improving water repellency and friability of tannin-furanic foams by oil-grafted flavonoid tannins. *Bioresources* **2016**, *11*, 7754–7768. [CrossRef]
80. Santiago, F.J.; Delgado, C.; Basso, M.C.; Pizzi, A.; Fierro, V.; Celzard, A. Mechanically blown wall-projected tannin-based foams. *Ind. Crops Prod.* **2018**, *113*, 316–323. [CrossRef]
81. Santiago, F.J.; Tenorio, A.; Delgado, C.; Basso, M.C.; Pizzi, A.; Fierro, V.; Celzard, A.; Sanchez, M.C.; Franco, J.M. Projectable tannin foams by mechanical and chemical expansion. *Ind. Crops Prod.* **2018**, *120*, 90–96. [CrossRef]
82. Li, X.; Pizzi, A.; Cangemi, M.; Fierro, V.; Celzard, A. Flexible natural tannin-based and protein-based biosourced foams. *Ind. Crops Prod.* **2012**, *37*, 389–393. [CrossRef]
83. Basso, M.C.; Giovando, S.; Pizzi, A.; Pasch, H.; Pretorius, N.; Delmotte, L.; Celzard, A. Flexible-Elastic copolymerized polyurethane-tannin foams. *J. Appl. Polym. Sci.* **2014**, *131*, 40499. [CrossRef]
84. Delgado, C.; Amaral, G.; Grishechko, L.; Sanchez, A.; Fierro, V.; Pizzi, A.; Celzard, A. Fire-resistant tannin-ethylene glycol gels working as rubber springs with tuneable elastic properties. *J. Mater. Chem. A* **2017**, *5*, 14720–14732. [CrossRef]
85. Lacoste, C.; Basso, M.C.; Pizzi, A.; Laborie, M.-P.; Celzard, A.; Fierro, V. Pine tannin-based rigid foams: Mechanical and thermal properties. *Ind. Crops Prod.* **2013**, *43*, 245–250. [CrossRef]
86. Lacoste, C.; Pizzi, A.; Basso, M.C.; Laborie, M.-P.; Celzard, A. Pinus pinaster tannin/furanic foams: Part 1, Formulation. *Ind. Crops Prod.* **2014**, *52*, 450–456. [CrossRef]
87. Tondi, G.; Pizzi, A.; Pasch, H.; Celzard, A. Structure degradation, conservation and rearrangement in the carbonization of polyflavonoid tannin/furanic rigid foams—A MALDI—TOF investigation. *Polym. Degrad. Stabil.* **2008**, *93*, 968–975. [CrossRef]
88. Celzard, A.; Szczekurek, A.; Jana, P.; Fierro, V.; Basso, M.C.; Bourbigot, S.; Stauber, M.; Pizzi, A. Latest progresses in tannin-based cellular solids. *J. Cell. Plast.* **2015**, *51*, 89–102. [CrossRef]
89. Lacoste, C.; Basso, M.C.; Pizzi, A.; Celzard, A.; Ella Bang, E.; Gallon, N.; Charrier, B. Pine (*Pinus pinaster*) and quebracho (*Schinopsis lorentzii*) tannin based foams as green acoustic absorbers. *Ind. Crops Prod.* **2015**, *67*, 70–73. [CrossRef]
90. Tondi, G.; Fierro, V.; Pizzi, A.; Celzard, A. Tannin-based carbon foams. *Carbon* **2009**, *47*, 1480–1492. [CrossRef]
91. Abdalla, S.M.S.; Al-Marzouki, F.; Pizzi, A.; Bahabri, F.S. Bone Graft with a Tannin-Hydroxyapatite Scaffold and Stem Cells for Bone Engineering. U.S. Patent 10,155,069, 18 December 2018.
92. Roux, D.G.; Ferreira, D.; Hundt, H.K.L.; Malan, E. Structure, stereochemistry and reactivity of condensed tannins as basis for their extended industrial application. *Appl. Polym. Symp.* **1975**, *28*, 335–353.
93. Palma, G.; Freer, J.; Baeza, J. Removal of metal ions by modified Pinus Radiata bark and tannins from water solutions. *Water Res.* **2003**, *37*, 4974–4980. [CrossRef] [PubMed]
94. Zhan, X.-M.; Zhao, X. Mechanism of lead adsorption from aqueous solutions using an adsorbent synthesized from natural condensed tannin. *Water Res.* **2003**, *37*, 3905–3912. [CrossRef]
95. Matsumura, T.; Usuda, S. Applicability of insoluble tannin to treatment of waste containing americium. *J. Alloys Comp.* **1998**, *271–273*, 244–247. [CrossRef]
96. Ogata, T.; Nakano, Y. Mechanisms of gold recovery from aqueous solutions using a novel tannin gel adsorbent synthesized from natural condensed tannin. *Water Res.* **2005**, *39*, 4281–4286. [CrossRef] [PubMed]
97. Yu, B.; Zhang, Y.; Shukla, A.; Shukla, S.S.; Dorris, K.L. The removal of heavy metals from aqueous solutions by sawdust adsorption - removal of lead and comparison of its adsorption with copper. *J. Hazard. Mat.* **2001**, *84*, 83–94. [CrossRef]

98. Sciban, M.; Klasnja, M. Wood sawdust and wood originate materials as adsorbents for heavy metal ions. *Holz als Roh-und Werkstoff* **2004**, *62*, 69–73. [CrossRef]
99. Nakajima, A.; Ohe, K.; Baba, Y.; Kijima, T. Mechanism of Gold Adsorption by Persimmon Tannin Gel. *Anal. Sci.* **2003**, *19*, 1075–1077. [CrossRef]
100. Vazquez, G.; Gonzalez-Alvarez, J.; Freire, S.; Lopez-Moren, M.; Antorrena, G. Removal of cadmium and mercury ions from aqueous solution by sorption on treated *Pinus pinaster* bark: Kinetics and isotherms. *Bioresour. Technol.* **2002**, *82*, 247–251. [CrossRef]
101. Yamaguchi, H.; Higashida, R.; Higuchi, M.; Sakata, I. Adsorption mechanism of heavy-metal ion by microspherical tannin resin. *J. Appl. Polym. Sci.* **1992**, *45*, 1463–1472. [CrossRef]
102. Tondi, G.; Oo, C.W.; Pizzi, A.; Trosa, A.; Thevenon, M.F. Metal adsorption of tannin-based rigid foams. *Ind. Crops Prod.* **2009**, *29*, 336–340. [CrossRef]
103. Oo, C.W.; Kassim, M.J.; Pizzi, A. Characterization and performance of *Rhizophora apiculata* mangrove polyflavonoid tannins in the adsorption of copper (II) and lead (II). *Ind. Crops Prod.* **2009**, *30*, 152–161. [CrossRef]
104. Bacelo, H.; Vieira, B.R.C.; Santos, S.C.R.; Boaventure, R.A.R.; Botelho, C.M.S. Recovery and valorization of tannins from a forest waste as an adsorbent for antimony uptake. *J. Clean. Prod.* **2018**, *198*, 1324–1355. [CrossRef]
105. Grenda, K.; Arnold, J.; Hunkeler, D.; Gamelas, J.A.F.; Rasteiro, M.G. Tannin-based coagulants from laboratory to pilot plant scales for coloured wastewater treatment. *Bioresources* **2018**, *13*, 2727–2747. [CrossRef]
106. Matamala, G.; Smeltzer, W.S.; Droguetta, G. Comparison of steel anticorrosive protection formulated with natural tannins extracted from acacia and from pine bark. *Corros. Sci.* **2000**, *42*, 1351–1362. [CrossRef]
107. Slabbert, N.E. Complexation of Condensed Tannins with Metal Ions. In *Plant Polyphenols*; Hemingway, R.W., Laks, P.E., Eds.; Basic Life Sciences; Springer: Boston, MA, USA, 1992; Volume 59, pp. 421–436.
108. Ghazali, N.A.; Naganawa, S.; Masuda, Y. Feasibility Study of Tannin-Lignosulfonate Drilling Fluid System for Drilling Geothermal Prospect. In Proceedings of the 43rd Workshop on Geothermal Reservoir Engineering, Stanford University, Stanford, CA, USA, 12–14 February 2018; p. 1.
109. Basso, M.C.; Pizzi, A.; Polesel-Maris, J.; Delmotte, L.; Colin, B.; Rogaume, Y. MALDI-TOF and 13C NMR analysis of the cross-linking reaction of condensed tannins by triethyl phosphate. *Ind. Crops Prod.* **2017**, *95*, 621–631. [CrossRef]
110. Polesel-Maris, J.; Joutang, I. Anti-Adhesive Coating Based on Condensed Tannins. Patent Convention Treaty (PCT) WO2017/037393A1, 9 March 2017.
111. Basso, M.C.; Pizzi, A.; Delmotte, L.; Abdalla, S. MALDI-TOF and 13C NMR Analysis of the Cross-linking Reaction of Lignin by Triethyl Phosphate. *Polymers* **2017**, *9*, 206. [CrossRef]
112. Smithers, V. Absorbent Materials for Floral Arrangements. U.S. Patent 2,753,277A, 3 July 1956.
113. Pilato, L. Floral foam product and method of producing the same which incorporates a flower preservative and a bactericide. U.S. Patent 4,225,679A, 30 September 1980.
114. Landrock, A.H. *Handbook of Plastic Foams: Types, Properties, Manufacture and Applications*; Noyes Data Corporation Publishers: Park Ridge, NJ, USA, 1995.
115. Pilato, L. *Phenolic Resins: A Century of Progress*; Springer: Bound Brook, NJ, USA, 2010.
116. Basso, M.C.; Pizzi, A.; Al-Marzouki, F.; Abdalla, S. Horticultural/hydroponics and floral foams from tannins. *Ind. Crops Prod.* **2016**, *87*, 177–181. [CrossRef]
117. McKenzie, A.E.; Yuritta, Y.P. Starch corrugating adhesives. *Appita* **1974**, *26*, 30–34.
118. Thebault, M.; Pizzi, A.; Dumarcay, S.; Gerardin, P.; Fredon, E.; Delmotte, L. Polyurethanes from hydrolydsable tannins obtained without using isocyanates. *Ind. Crops Prod.* **2014**, *59*, 329–336. [CrossRef]
119. Thebault, M.; Pizzi, A.; Essawy, H.; Baroum, A.; van Assche, G. Isocyanate free condensed tannin-based polyurethanes. *Eur. Polym. J.* **2015**, *67*, 513–523. [CrossRef]
120. Thebault, M.; Pizzi, A.; Santiago, F.J.; Al-Marzouki, F.M.; Abdalla, S. Isocyanate-free polyurethanes by coreaction of condensed tannins with aminated tannins. *J. Renew. Mater.* **2017**, *5*, 21–29. [CrossRef]
121. Xi, X.; Pizzi, A. Delmotte, Isocyanate-free Polyurethane Coatings and Adhesives from Mono- and Di-Saccharides. *Polymers* **2018**, *10*, 402. [CrossRef] [PubMed]
122. Li, X.; Nicollin, A.; Pizzi, A.; Zhou, X.; Sauget, A.; Delmotte, L. Natural tannin/furanic thermosetting moulding plastics. *RSC Adv.* **2013**, *3*, 17732–17740. [CrossRef]

123. Lagel, M.C.; Pizzi, A.; Basso, M.C.; Abdalla, S. Development and characterisation of abrasive grinding wheels with a tannin-furanic resin matrix. *Ind. Crops Prod.* **2014**, *65*, 333–348.
124. Lagel, M.C.; Hai, L.; Pizzi, A.; Basso, M.C.; Delmotte, L.; Abdalla, S.; Zahed, A.; Al-Marzouki, F.M. Automotive brake pads made with a bioresin matrix. *Ind. Crops Prod.* **2015**, *85*, 372–381. [CrossRef]
125. Al-Marzouki, F.M.; Zahed, A.; Pizzi, A.; Abdalla, S.M.S. Thermosetting Resin Composition for Brake Pads, Method of Preparation and Brake Pad Assembly. U.S. patent 9,791,012 B1, 17 October 2017.
126. Boutevin, B.; Caillol, S.; Burguiere, C.; Rapior, S.; Fulcrand, H.; Nouailhas, H. Novel Method for Producing Thermosetting Epoxy Resins. Patent Convention Treaty WO2010136725A1, 2 December 2010.
127. Nouailhas, H.; Aouf, C.; Guerneve, C.; Caillol, S.; Boutevin, B.; Fulcrand, H. Synthesis and Properties of Biobased Epoxy Resins. Part 1. Glycidylation of Flavonoids by Epichlorohydrin. *J. Polym. Sci. Part A Polym. Chem.* **2011**, *49*, 2261–2270. [CrossRef]
128. Jahanshahi, S.; Pizzi, A.; Abdolkhani, A.; Doosthoseini, K.; Shakeri, A.; Lagel, M.C.; Delmotte, L. MALDI-TOF and 13C-NMR and FT-MIR and strength characterization of glycidyl ether tannin epoxy resins. *Ind. Crops Prod.* **2016**, *83*, 177–185. [CrossRef]
129. Jahanshahi, S.; Pizzi, A.; Abdolkhani, A.; Shakeri, A. Analysis and testing of bisphenol-A-free bio-based tannin epoxy-acrylic adhesives. *Polymers* **2016**, *8*, 143. [CrossRef]
130. Tomita, H.; Yonezawa, K. Epoxy Resin and Process to Prepare the Same. Europeran Patent EP0095609B1, 7 December 1983.
131. Aouf, C.; La Guerneve, S.; Caillol, S.; Fulcrand, H. Study of the O-glycidylation of natural phenolic compounds. The relationship between the phenolic structure and the reaction mechanism. *Tetrahedron* **2013**, *69*, 1345–1353. [CrossRef]
132. Benyahya, S.; Aouf, C.; Caillol, S.; Boutevin, B.; Fulcrand, H. Functionalized green tea tannins as phenolic prepolymers for bio-based epoxy resins. *Ind. Crops Prod.* **2014**, *53*, 296–307. [CrossRef]
133. Pizzi, A.; Kueny, R.; Lecoanet, F.; Masseteau, B.; Carpentier, D.; Krebbs, A.; Loiseau, F.; Molina, S.; Ragoubi, M. High resin content natural matrix-natural fibre biocomposites. *Ind. Crops Prod.* **2009**, *30*, 235–240. [CrossRef]
134. Nicollin, A.; Kueny, R.; Toniazzo, L.; Pizzi, A. High density composites from natural fibres and tannin resin. *J. Adhes. Sci. Technol.* **2012**, *26*, 1537–1545.
135. Sauget, A.; Nicollin, A.; Pizzi, A. Fabrication and mechanical analysis of mimosa tannin and commercial flax fibers biocomposites. *J. Adhes. Sci. Technol.* **2013**, *27*, 2204–2218. [CrossRef]
136. Nicollin, A.; Li, X.; Girods, P.; Pizzi, A.; Rogaume, Y. Fast pressing composite using tannin-furfuryl alcohol resin and vegetal fibers reinforcement. *J. Renew. Mater.* **2013**, *1*, 311–316. [CrossRef]
137. Abdullah, U.H.B.; Pizzi, A.; Zhou, X.; Rode, K.; Delmotte, L. Mimosa tannin resins for impregnated paper overlays. *Eur. J. Wood Wood Prods.* **2013**, *71*, 153–162. [CrossRef]
138. Abdullah, U.H.B.; Pizzi, A.; Zhou, X.; Merlin, A. A note on the surface quality of plywood overlaid with mimosa (Acacia mearnsii) tannin and melamine urea formaldehyde impregnated paper: Effects of moisture content of resin-impregnated papers before pressing on the physical properties of overlaid panels. *Int. Wood Prod. J.* **2013**, *4*, 253–256. [CrossRef]
139. Abdullah, U.H.B.; Pizzi, A.; Zhou, X. High pressure paper laminates from mimosa tannin resin. *Int. Wood Prod. J.* **2014**, *5*, 224–227. [CrossRef]
140. Basso, M.C.; Lacoste, C.; Pizzi, A.; Fredon, E.; Delmotte, L. Flexible tannin-furanic films and lacquers. *Ind. Crops Prod.* **2014**, *61*, 352–360. [CrossRef]
141. Braghiroli, F.; Fierro, V.; Pizzi, A.; Rode, K.; Radke, W.; Delmotte, L.; Parmentier, J.; Celzard, A. Condensation reaction of flavonoid tannins with ammonia. *Ind. Crops Prod.* **2013**, *44*, 330–335. [CrossRef]
142. Hewitt, P.C. *Cement Admixtures: Uses and Applications*, 2nd ed.; Longman; The Bath Press: London, UK, 1988; Chapter 1.
143. Khalil, S.M. Influence of a lignin based admixture on the hydration of portland cements. *Cement Concr. Res.* **1973**, *3*, 677–688. [CrossRef]
144. Hewitt, P.C. *Cement Admixtures: Uses and Applications*, 2nd ed.; Longman; The Bath Press: London, UK, 1988; Chapter 7.

145. Kaspar, H.R.E.; Pizzi, A. Industrial plasticizing/dispersion aids for cement based on polyflavonoid tannins. *J. Appl. Polym. Sci.* **1996**, *59*, 1181–1190. [CrossRef]

 © 2019 by the author. Licensee MDPI, Basel, Switzerland. This article is an open access article distributed under the terms and conditions of the Creative Commons Attribution (CC BY) license (http://creativecommons.org/licenses/by/4.0/).

MDPI
St. Alban-Anlage 66
4052 Basel
Switzerland
Tel. +41 61 683 77 34
Fax +41 61 302 89 18
www.mdpi.com

Biomolecules Editorial Office
E-mail: biomolecules@mdpi.com
www.mdpi.com/journal/biomolecules

www.ingramcontent.com/pod-product-compliance
Lightning Source LLC
LaVergne TN
LVHW070607100526
838202LV00012B/589